AF411312

Organic-Inorganic Hybrid Materials II

Organic-Inorganic Hybrid Materials II

Editors

Jesús-María García-Martínez
Emilia P. Collar

MDPI • Basel • Beijing • Wuhan • Barcelona • Belgrade • Manchester • Tokyo • Cluj • Tianjin

Editors
Jesús-María García-Martínez
Polymer Engineering Group
(GIP)-Institute of Polymer
Science and Technology
(ICTP)
CSIC
Madrid
Spain

Emilia P. Collar
Polymer Engineering Group
(GIP)-Institute of Polymer
Science and Technology
(ICTP)
CSIC
Madrid
Spain

Editorial Office
MDPI
St. Alban-Anlage 66
4052 Basel, Switzerland

This is a reprint of articles from the Special Issue published online in the open access journal *Polymers* (ISSN 2073-4360) (available at: www.mdpi.com/journal/polymers/special_issues/ Organic_Inorganic_Hybrid_II).

For citation purposes, cite each article independently as indicated on the article page online and as indicated below:

LastName, A.A.; LastName, B.B.; LastName, C.C. Article Title. *Journal Name* **Year**, *Volume Number*, Page Range.

ISBN 978-3-0365-7079-2 (Hbk)
ISBN 978-3-0365-7078-5 (PDF)

Cover image courtesy of Jesús-María García-Martínez and Emilia P. Collar

Contents

About the Editors

Jesús-María García-Martínez

Jesús-María García-Martínez holds a Ph.D. in Chemistry (Chemical Engineering) from Universidad Complutense de Madrid (1995), two M.Sc. level degrees, and has received more than 70 highly specialized courses. He is a Tenured Scientist at the Institute of Polymer Science and Technology (ICTP) of the Spanish National Research Council (CSIC). Since 1992, within the Polymer Engineering Group (GIP), he has co-authored more than 175 scientific and/or technical works in topics related to polymer engineering, chemical modification of polymers, heterogeneous materials based on polymers, interphases, and interfaces, polymer composites blends and alloys, organic-inorganic hybrids materials, polymer recycling, quality, and so on. Furthermore, he has participated in more than 32 research and industrial projects (national and international programs) and is co-author of one currently active industrial patent on polymer recycling. From 2000–2005, he also assumed the position of Quality Director for the ICTP (CSIC) ISO 17025 Accreditation Project of the ICTP Laboratories. Additionally, Dr. García-Martínez is actively reviewing tasks for WOS and SCOPUS indexed journals, with more than 270 reports in the last years, being awarded twice with the Publons Reviewer Awards (2018, 2019) and once with the POLYMERS Outstanding Reviewer Award (2019). Since 2016, he has served as Head of the Department of Chemistry and Properties of Polymer Materials within the ICTP (CSIC).

Emilia P. Collar

Ph.D. in Industrial Chemistry (U. Complutense, 1986). Since 1990 is a permanent staff (Tenured Scientist) at the Consejo Superior de Investigaciones Científicas (CSIC), after two years (1986–88) as CSIC's postdoc fellow and one (1989) as Chemical Engineering Assistant Teacher at the Universidad Complutense de Madrid. In the Polymer Science and Technology Institute (ICTP/CSIC), she works at the Polymer Engineering Group (GIP), founded in 1982 by Prof. O. Laguna, being the GIP's Head since 1999. Between 1990 and 2005, she supervised six doctoral theses and five post-graduate ones, under 14 research public funds projects, jointly to 18 research private funded contracts issuing 53 technical reports for different companies. Furthermore, from 2001 to 2005, she performed the positions of Technical Director of the Physical and Mechanical Properties Laboratory and Deputy-Technical Director of the Thermal Properties Laboratory under the successful ISO 17025 Accreditation Project for the CSIC/ICTP's Laboratories, ACiTP, commanded by the ICTP's Head. Author and co-author of more than 20 chapters on books, two currently active industrial patents on polymer recycling, and more than 150 papers mainly on SCI Journals, her research work lines deal with polymers and environment under the general frame of heterogeneous materials based on polymers. From 2006 to date, she has participated in four public Spanish-funded research projects and one EU project under its 7th Framework Program.

Preface to "Organic-Inorganic Hybrid Materials II"

Let us introduce this new reprint published by MDPI. It is well worth mentioning that this volume is the second part of a previous one devoted to Organic-Inorganic Hybrid Materials, recently published by MDPI in 2021. This new one includes ten new original approaches to the topic.

As expected, these unique contributions fully agree with the definition of hybrid material as per the IUPAC (International Union of Pure and Applied Chemistry) recommendations, defining hybrid materials as the composed of an intimate mixture of inorganic or both types of components interpenetrating scales of less than one micron. Therefore, the reader may deduce that this definition offers many approximations to investigate in the field.

Complementary, the literature usually classifies the organic-inorganic hybrid systems into two families or classes (Class-I and Class-II), attending to the type of interactions between the phases (weak or strong, respectively). So, Class-I organic-inorganic hybrid compounds imply weak interactions such as Van der Waals, hydrogen bonds, electro-static, etc. Conversely, Class-II compounds mean the existence of chemical bonds between phases.

Additionally, a third option is possible, combining the two types of interactions in the same hybrid system.

In essence, we can conclude that the organic-inorganic materials are multi-component systems with at least one of their organic (the polymer) or inorganic components in the nano-metric size domain, which confers the material as a whole greatly enhanced properties respecting the constitutive parts in isolation.

The works compiled in this volume are at the forefront of Organic–Inorganic Hybrid material research. So, these ten new articles would provide information and inspiration for future applications in this fantastic and fascinating scientific field.

Jesús-María García-Martínez and Emilia P. Collar
Editors

Editorial

Organic–Inorganic Hybrid Materials II: Some Additional Contributions to the Topic

Jesús-María García-Martínez * and Emilia P. Collar *

Polymer Engineering Group (GIP), Polymer Science and Technology Institute (ICTP),
Spanish Council for Scientific Research (CSIC), C/Juan de la Cierva, 3, 28006 Madrid, Spain
* Correspondence: jesus.maria@ictp.csic.es (J.-M.G.-M.); ecollar@ictp.csic.es (E.P.C.)

Citation: García-Martínez, J.-M.;
Collar, E.P. Organic–Inorganic Hybrid
Materials II: Some Additional
Contributions to the Topic. *Polymers*
2021, 13, 2390. https://doi.org/
10.3390/polym13152390

Received: 19 July 2021
Accepted: 20 July 2021
Published: 21 July 2021

Publisher's Note: MDPI stays neutral
with regard to jurisdictional claims in
published maps and institutional affiliations.

By following the successful editorial pathway of the recently published former Special Issue dedicated to Organic–Inorganic Hybrid Materials [1,2], this second part offers a number of new original contributions and/or approaches to the topic. Once again, these new investigations matched the definition of hybrid material as per the IUPAC (International Union of Pure and Applied Chemistry) recommendations [3]. That means a hybrid material is composed of an intimate mixture of inorganic components, organic components, or both types of component, which usually interpenetrate on scales of less than 1 µm [3].

The latter implies that this definition can be identified with a myriad of industrial and/or academic approaches. In fact, when applied to organic–inorganic hybrid materials open plenty of research and design lines for material scientists. It is noteworthy that the organic–inorganic hybrid materials are multi-component compounds with at least one of the parts in the sub-micrometric and/or the nano-metric domain [4].

Additionally, the literature usually classifies the organic–inorganic hybrid systems into two families or classes (Class-I and Class-II), attending to the type of interactions between the phases (weak or strong, respectively) [5]. Therefore, Class-I organic–inorganic hybrid compounds imply weak interactions such as Van der Waals, hydrogen bonds, electrostatic, etc.; conversely, Class-II compounds have a real chemical bond between phases. The third option is a combination of the two types of interactions in the same hybrid system [6]. In essence, it can be said that the organic–inorganic materials are multi-component compounds with at least one of their organic (the polymer) or inorganic components in the nano-metric-size domain, which confers the material as a whole of greatly enhanced properties respecting the constitutive parts in isolation [4–7].

Under these premises, this additional Special Issue complements the previous one [1,2] with five new approaches, covering a wide range of conceptual and application spectra.

Therefore, the first article by Figueira et al. [8], dedicated to the memory of one of the co-authors (Professor Carlos. J. R. Silva), is devoted to the study of the synthesis and characterization of five new organic–inorganic hybrids (OIH) sol-gel materials obtained from a functionalized siloxane 3-glycidoxypropyltrimethoxysilane (GPTMS) by the reaction with a new Jeffamine®, a secondary diamine, (SD-2001), and a triamine, (T-403). The resulting materials (OIH sol-gel materials) were studied and characterized by UV-visible absorption spectro-photometry, steady-state photoluminescence spectroscopy, and electrochemical impedance spectroscopy. From there, it was found that the optical and electrical properties of the obtained new OIH materials exhibited promising properties as support films in an optical sensor area. Additionally, some investigations were developed on the chemical stability of the OIH materials in contact with cement pastes after three different periods of time, concluding that many of the new materials exhibited improved behaviors which were very promising in the alkaline environments caused by the cement paste. This fact makes these materials to very promising candidates for their potential application as a support film in optical fiber sensors in the area of civil engineering.

The work by Alsame et al. [9] is focused on an ease hydrothermal synthesis of polyaniline (PANI)-modified molybdenum disulfide (MoS2) nanosheets employed to obtain a novel 3D organic–inorganic hybrid material. An exhaustive characterization of the obtained material by using field emission scanning electron microscopy, high-resolution transmission electron microscopy, energy-dispersive X-ray spectroscopy and X-ray diffraction studies led the authors to conclude that the obtaining procedure of the organic–inorganic hybrid material (PANI–MoS$_2$) was successful. Additionally, this organic–inorganic hybrid material has demonstrated its efficiency in the extraction and pre-concentration of trace mercury ions, thanks to the selective complexation between the sulfur ion of PANI–MoS$_2$ with Hg(II)—the lowest concentration threshold being as little as 0.03 µg L^{-1}—being applied to the proper analysis of real samples. This suggests that the PANI–MoS$_2$ hybrid material can be used for trace Hg(II) analyses for environmental water monitoring.

The design of a flat-sheet direct-contact membrane distillation (DCMD) module to remove nitrates from water is the purpose of the investigation by Orroji, Razmjou, and Ebrahimi [10]. For such a purpose, a polyvinylidene fluoride (PVDF) membrane was employed in a DCMD process performed at 1 atm and below 100 °C. It was found that, during the entire process, the electrical conductivity of the feed containing nitrate increased, while the permeation remained constant, resulting in the fact that the nitrate ions did not pass through the membrane while pure water is obtained. In order to have a better hierarchical surface control, the authors proposed modifying the membrane with TiO$_2$ nano-particles for multi-layer roughness on the micro/nano-scale domains. Further, the incorporation of 1H,1H,2H,2H-Perfluorododecyltrichlorosilane (FTCS) to the TiO$_2$-modified surface of the membrane with the purpose to obtain super-hydrophobic properties improving its separation performance, was also explored. Finally, the authors compared types of membranes by the experimental, pilot-scale, and simulation procedures, and found very interesting results and predictions about the surface needed for a proper separation process.

The work contributed by Ahmad et al. [11] focused on the kinetics of photo-isomerization and the time evolution of hybrid thin films, but with a novel insight approach based on considering the azo-dye methyl red (MR) group linked into graphene embedded onto a polyethylene oxide matrix (PEO), forming an organic–inorganic hybrid material, PEO-(MR-Graphene). The kinetics study was undertaken by UV-Vis and FTIR spectroscopy. Additionally, they used new models developed with new analytical methods. The authors reported that the presence of azo-dye MR in the compound is the key aspect for the resource action of the trans to cis cycles (and vice versa) through UV to visible-illumination relaxations. Consequently, the authors found that these hybrid composite thin films may be used as applicants for photochromic molecular switches, light-gated transistors, and molecular solar thermal energy storage media.

Finally, and because the properties of hybrid silica xerogels obtained by the sol-gel method are strongly dependent on both the precursor and the synthesis conditions, Espinal-Viguri and Garrido et al. [12] investigate the influence of the organic substituent of the precursor on the sol-gel process conditioning the final structure of the obtained materials. The latter applies to the xerogels with tetraethyl orthosilicate (TEOS) and alkyltriethoxysilane or chloroalkyltriethoxysilane at different molar percentages (RTEOS and ClRTEOS, R = methyl [M], ethyl [E], or propyl [P]). The authors conclude that the intermolecular forces caused by the organic groups jointly to the chlorine atom in the precursors were ascertained by comparison of the sol-gel process between alkyl and chloroalkyl series. Therefore, the authors conclude that the alkyl chain and the chlorine atom of the precursor in these materials determine their inductive and steric effects on the sol-gel process. Additionally, the authors studied the microstructure of the obtained xerogels by deconvolution methods to Fourier-transformed infrared spectra. Therefore, the distribution of (SiO)$_4$ and (SiO)$_6$ rings into the silicon matrix of the xerogels is revealed. These results are fully consistent with X-ray diffraction spectra, indicating that the local periodicity associated with four-fold rings increases the amount of the precursor. To conclude, the authors found that the combination of the sol-gel process and the emerging ordered domains determine the final structure of the hybrid materials and, therefore, their properties and potential applications.

To summarize, in the light of the present and former [1,2] Special Issues, it can be affirmed that the topic of organic–inorganic hybrid materials covers a broad spectrum of approaches and/or applications. In any case, it is the inter-phase between the components which plays a crucial role in these final performances of the hybrid materials. This fact becomes the critical aspect when intend to obtain this type of hybrid materials with "tailor-made" organized structures in every one of the nano-, meso-, micro- and meso-scale factors ultimately involved.

We would like to thank all the authors for their contributions and encourage them to continue enthusiastically with their research.

Funding: This research received no external funding.

Conflicts of Interest: The authors declare no conflict of interest.

References

1. García-Martínez, J.M.; Collar, E.P. Organic–Inorganic Hybrid Materials. *Polymers* **2021**, *13*, 86. [CrossRef]
2. García-Martínez, J.M.; Collar, E.P. (Eds.) *Organic–Inorganic Hybrid Materials*; MDPI Books: Basel, Swizertland, 2021; ISBN1 978-3-0365-1301-0 (Hbk). ISBN2 978-3-0365-1302-7 (PDF).
3. Alemán, J.; Chadwick, A.V.; He, J.; Hess, M.; Horie, K.; Jones, R.G.; Kratochvíl, P.; Meisel, I.; Mita, I.; Moad, G.; et al. Definitions of terms relating to the structure and processing of sols, gels, networks, and inorganic–organic hybrid materials (IUPAC recommendations 2007). *Pure Appl. Chem.* **2007**, *79*, 1801–1829. [CrossRef]
4. Pielichowski, K.; Majka, T.M. *Polymer Composites with Functionalized Nanoparticles: Synthesis, Properties, and Applications*; Elsevier Inc.: Amsterdam, The Netherlands, 2019; pp. 1–504.
5. Faustini, M.; Nicole, L.; Ruiz-Hitzky, E.; Sanchez, C. History of Organic-Inorganic Hybrid Materials: Prehistory, Art, Science, and Advanced Applications. *Adv. Funct. Mater.* **2018**, *28*, 1704158. [CrossRef]
6. Pogrebnjak, A.D.; Beresnev, V.M. *Nanocoatings Nanosystems Nanotechnologies*; Bentham Books: Sharjah, United Arab Emirates, 2012.
7. Collar, E.P.; Areso, S.; Taranco, J.; García-Martínez, J.M. Heterogeneous Materials based on Polypropylene. In *Polyolefin Blends*, 1st ed.; Nwabunma, D., Kyu, T., Eds.; Wiley-Interscience: Hoboken, NJ, USA, 2008; pp. 379–410.
8. Gomes, B.R.; Figueira, R.B.; Costa, S.P.G.; Raposo, M.M.M.; Silva, C.J.R. Synthesis, Optical and Electrical Characterization of Amino-Alcohol Based Sol-Gel Hybrid Materials. *Polymers* **2020**, *12*, 2671. [CrossRef] [PubMed]
9. Ahmad, H.; BinSharfan, I.I.; Khan, R.A.; Alsalme, A. 3D Nanoarchitecture of Polyaniline-MoS$_2$ Hybrid Material for Hg(II) Adsorption Properties. *Polymers* **2020**, *12*, 2731. [CrossRef] [PubMed]
10. Ebrahimi, F.; Orooji, Y.; Razmjou, A. Applying Membrane Distillation for the Recovery of Nitrate from Saline Water Using PVDF Membranes Modified as Superhydrophobic Membranes. *Polymers* **2020**, *12*, 2774. [CrossRef]
11. Al-Bataineh, Q.M.; Ahmad, A.A.; Alsaad, A.M.; Telfah, A. New Insight on Photoisomerization Kinetics of Photo-Switchable Thin Films Based on Azobenzene/Graphene Hybrid Additives in Polyethylene Oxide. *Polymers* **2020**, *12*, 2954. [CrossRef]
12. Cruz-Quesada, G.; Espinal-Viguri, M.; López-Ramón, M.V.; Garrido, J.J. Hybrid Xerogels: Study of the Sol-Gel Process and Local Structure by Vibrational Spectroscopy. *Polymers* **2021**, *13*, 2082. [CrossRef] [PubMed]

 *polymers*

Article

Unmodified Silica Nanoparticles Enhance Mechanical Properties and Welding Ability of Epoxy Thermosets with Tunable Vitrimer Matrix

Anna I. Barabanova [1,*], Egor S. Afanas'ev [1], Vyacheslav S. Molchanov [2], Andrey A. Askadskii [1,3] and Olga E. Philippova [2]

[1] A.N. Nesmeyanov Institute of Organoelement Compounds, Russian Academy of Sciences, 119991 Moscow, Russia; nambrot@yandex.ru (E.S.A.); andrey@ineos.ac.ru (A.A.A.)
[2] Physics Department, Moscow State University, 119991 Moscow, Russia; molchan@polly.phys.msu.ru (V.S.M.); phil@polly.phys.msu.ru (O.E.P.)
[3] Moscow State University of Civil Engineering, 129337 Moscow, Russia
* Correspondence: barabanova@polly.phys.msu.ru

Abstract: Epoxy/silica thermosets with tunable matrix (vitrimers) were prepared by thermal curing of diglycidyl ether of bisphenol A (DGEBA) in the presence of a hardener—4-methylhexahydrophthalic anhydride (MHHPA), a transesterification catalyst—zinc acetylacetonate (ZAA), and 10–15 nm spherical silica nanoparticles. The properties of the resulting material were studied by tensile testing, thermomechanical and dynamic mechanical analysis. It is shown that at room temperature the introduction of 5–10 wt% of silica nanoparticles in the vitrimer matrix strengthens the material leading to the increase of the elastic modulus by 44% and the tensile stress by 25%. Simultaneously, nanoparticles enhance the dimensional stability of the material since they reduce the coefficient of thermal expansion. At the same time, the transesterification catalyst provides the thermoset with the welding ability at heating, when the chain exchange reactions are accelerated. For the first time, it was shown that the silica nanoparticles strengthen welding joints in vitrimers, which is extremely important, since it allows to repeatedly use products made of thermosets and heal defects in them. Such materials hold great promise for use in durable protective coatings, adhesives, sealants and many other applications.

Keywords: vitrimer; epoxy network; nanocomposite; silica nanoparticles

Citation: Barabanova, A.I.; Afanas'ev, E.S.; Molchanov, V.S.; Askadskii, A.A.; Philippova, O.E. Unmodified Silica Nanoparticles Enhance Mechanical Properties and Welding Ability of Epoxy Thermosets with Tunable Vitrimer Matrix. *Polymers* **2021**, *13*, 3040. https://doi.org/10.3390/polym13183040

Academic Editors: Jesús-María García-Martínez and Emilia P. Collar

Received: 30 July 2021
Accepted: 30 August 2021
Published: 9 September 2021

Publisher's Note: MDPI stays neutral with regard to jurisdictional claims in published maps and institutional affiliations.

1. Introduction

Vitrimers represent a new class of polymer materials first proposed 10 years ago by L. Leibler and co-workers [1–3]. They consist of covalent adaptable networks that can rearrange their topology via exchange reactions while keeping the cross-link density unchanged. The first vitrimers were based on epoxy networks undergoing transesterification reactions [1,3,4]. Later, many other exchange reactions were proposed [2,5] including transamination of vinylogous urethanes [2,5] transalkylation of triazolium salts [6], olefin metathesis [7,8], thermo-activated disulfide rearrangements [9] and others.

Vitrimers, like other thermosets, possess excellent mechanical properties, enhanced chemical resistance and thermostability [10]. At the same time, the exchange reactions provide to vitrimers some additional features, such as the ability for reprocessing, recycling, reshaping and healing, which are quite important both from economic and ecological points of view. Moreover, exchange reactions enable the surface welding of vitrimers [1,11–13]. Welding being a viable joining technique gives the opportunity to produce multilayered structures combining different materials with desirable properties. Before the invention of vitrimers, the welding was restricted only to thermoplastics, which can be welded, when they are softened (e.g., by heating above their melting point). Vitrimers opened the way to weld thermosets due to exchange reactions that provided high enough interfacial diffusion of polymer chains [12] in order to form the welded joint.

Addition of nanoparticles is a promising way to improve the properties of vitrimers. One can expect that nanoparticles possessing a large surface area to interact with polymer chains are able not only to enhance the mechanical properties and the resistance to weathering [14,15] of vitrimers, but also to facilitate the welding through the intermolecular interactions. However, by now, there are only few papers describing vitrimer nanocomposites. They concern mainly vitrimers with epoxy matrix. This is not unexpected because among different vitrimers, epoxy thermosets are of particular interest [16], since epoxy resins are the most used thermoset materials [17], which share ~70% of the global thermosets market [4]. They possess many valuable properties such as high adhesive strength, good chemical resistance, and low curing shrinkage [16]. The epoxy vitrimer nanocomposites studied by now were filled with silica [18], graphene oxide [19], cellulose nanocrystals [20], carbon black [21], carbon fiber [22] or carbon nanotubes [11]. It was shown that nanoparticles can increase the elastic modulus [18,19] and enhance the self-healing and self-repairing ability of vitrimers [19].

Among different fillers, silica nanoparticles have many advantages including low cost, reach possibilities to modify their surface functionality, and easily tunable size. However, to the best of our knowledge, there is only one paper devoted to epoxy/silica vitrimer nanocomposites [18]. It concerns so-called soft nanocomposites with rather low T_g. No studies of hard epoxy vitrimer nanocomposites with strongly cross-linked matrix are available by now. At the same time, such nanocomposites are very promising for various applications including durable protective coatings, adhesives, paints, sealants, windmill blades and so forth [16,23–25]. For instance, they are required in civil infrastructure applications as adhesives for joining composite materials or repairing civil structures [15].

Therefore, in this article, we studied hard silica nanocomposite thermosets made from an epoxy-based vitrimer. The vitrimer was prepared by curing one of the most extensively used commercial epoxy resins [26]—bisphenol A diglycidyl ether (DGEBA) with 4-methylhexahydrophthalic anhydride (MHHPA) in the presence of a transesterification catalyst zinc acetylacetonate (ZAA). The topology rearrangements in the vitrimer thus obtained result from thermoactivated and catalyzed transesterification reactions. We demonstrate that the mechanical properties of hard epoxy-based vitrimers can be reinforced with 10–15 nm spherical silica nanoparticles without preventing the topology rearrangements. Moreover, the nanofiller strengthens the welding joints, which opens wide perspectives for the development of welding of thermoset vitrimers allowing repairing, reprocessing or combining them with other materials.

2. Materials and Methods

2.1. Materials

Monomer DGEBA, transesterification catalyst ZAA from Sigma-Aldrich (Steinheim am Albuch, Baden-Wurttemberg, Germany) and curing agent MHHPA (purity 98%) from Acrus Organics (Geel, Antwerp, Belgian) were used without further purification. Their chemical structures are presented in Figure 1. Silica nanoparticles in the form of 30–31 wt% colloidal dispersion in methyl ethyl ketone (MEK-ST) were purchased in Nissan Chemical Corporation (Santa Clara, CA, USA) and used as received. The size of silica particles provided by the supplier is 10–15 nm.

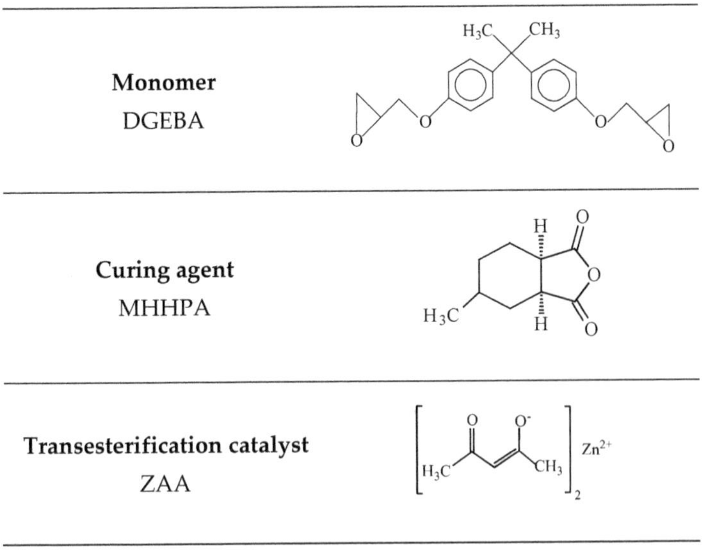

Monomer

DGEBA

Curing agent

MHHPA

Transesterification catalyst

ZAA

Figure 1. Chemical structure of the monomer bisphenol A diglycidyl ether DGEBA, curing agent 4-methylhexahydrophthalic anhydride MHHPA, and transesterification catalyst zinc acetylacetonate ZAA.

The content of surface silanol Si–OH groups was estimated by two complementary techniques: potentiometric titration and volumetric method. The first method is based on potentiometric titration of excess sodium hydroxide after its interaction with Si–OH groups on the surface of nanoparticles for 1 day with stirring [27]. The volumetric method is based on measurement of volume of gaseous products of reaction of silanol groups Si–OH with Grignard reagent (CH_3MgI in diisoamyl ether) [28]. The number of silanol groups estimated by the volumetric method was equal to 0.45 mmol/g, which is close to the value obtained by titration (0.5 mmol/g).

2.2. Synthesis of Epoxy/Silica Thermosets with Vitrimer Matrix

Samples of epoxy/silica thermosets with vitrimer matrix were prepared according to the method developed in [29]. ZAA powder (0.2134 g) was added to DGEBA (2.7866 g), and the mixture was homogenized at 150 °C for 3 h under stirring. Then, the SiO_2 colloid dispersion MEK-ST (0.9912 g) was added and the resulting mixture was heated at 80 °C for 2 h. After evaporation of the organic solvent MHHPA (2.6879 g) was added, the reaction mixture was stirred at room temperature and poured into a Teflon mold mounted on a horizontal plate in a Binder drying oven. The temperature was gradually increased (5–6 °C/min) to 140 °C and the reaction mixture was kept at this temperature for 12 h. The conditions of preparation of epoxy/silica networks with vitrimer matrices are given in Table 1.

Table 1. Conditions of preparation and mechanical properties of epoxy/silica vitrimer nanocomposites.

Sample	[DGEBA]/[MHHPA]/[ZAA], mol/mol/mol	$[SiO_2]$, wt%	Tensile Stress, MPa	Tensile Strain, %	Elastic Modulus, GPa	
					Tensile Tests	Guth's Prediction
V1	0.5/1.0/0.05	0	48	7.3	1.8	
N1-5	0.5/1.0/0.05	5	57	5.6	2.4	2.0
N1-10	0.5/1.0/0.05	10	60	4.9	2.6	2.2
V2	0.72/1.0/0.043	0	54	5.6	2.0	
N2-5	0.76/1.0/0.043	5	65	6.5	2.3	2.2
N2-10	0.76/1.0/0.04	10	68	6.5	2.4	2.3

The degree of curing α of the epoxy networks was determined using differential scanning calorimetry data obtained on a NETZCH DSC 204F1 instrument (Netzsch, Selb, Bayern, Germany) and calculated using the formula [30]:

$$\alpha = \frac{\Delta H_T - \Delta H_R}{\Delta H_T} \times 100\% \tag{1}$$

where ΔH_T is the amount of heat released during the nonisothermal DSC scanning of the initial reaction mixture and ΔH_R is the residual amount of heat released during the DSC scanning of the cured sample.

2.3. Mechanical Properties

Tensile tests of the native samples were performed on Lloyd LS5, AMETEK STC (Lloyd Instruments Ltd., Segensworth, Hampshire, UK) tensile machine at room temperature. Stress–strain curves were obtained on rectangular samples ($1.8 \times 4 \times 37$ mm) with a 5 kN load cell at constant cross-head speed of 2 mm/min and effective gauge length of 16 mm. The elastic modulus E was determined from the slope of the initial linear section of elastic deformation ε which the material undergoes under loading: $E = \sigma/\varepsilon$.

2.4. Thermomechanical Analysis

The thermomechanical analysis (TMA) of epoxy thermosets was performed on a TMA Q400 instrument (TA Industries, Woodland, CA, USA). Thermomechanical curves were recorded at a constant load of 100 g in the penetration mode using the probe with a diameter of 2.54 mm during heating in air from room temperature to 350 °C at a constant speed of 5 °C/min. The T_g value was determined as the mid-point of the temperature range of the transition of the sample from the glassy to rubbery state. The coefficient of linear thermal expansion (CTE) was measured in the expansion mode from slopes of the TMA curve before and above T_g. In the experiments, the samples were first heated to 250 °C and kept at this temperature for 10 min, then cooled to 25 °C, and again heated to 250 °C at a rate of 5 °C/min. The values of T_g and CTE were determined from the second heating cycle [31] to avoid the influence of the thermal and mechanical history of epoxy thermosets on their thermal properties.

The values of T_g, storage E', and loss E'' moduli of cured samples were determined with a dynamic mechanical analyzer NETZCH DMA 242 E Artemis (Netzsch, Selb, Bayern, Germany). The samples were cut to the size of $30 \times 6 \times 2$ mm before being mounted on a single cantilever clamp and measured at a frequency of 1.0 Hz and a heating rate of 1 °C/min from 25 to 250 °C. The storage modulus, loss modulus and loss factor, tan δ, were calculated as a function of temperature over a range from 25 to 250 °C. The glass transition temperature T_g was determined as a temperature corresponding to the peak value of tan δ.

2.5. Fourier-Transform Infrared Spectroscopy (FTIR)

FTIR spectroscopy measurements were carried out on Bruker Vertex 70v FTIR spectrometer (Bruker Optic GmbH, Ettlingen, Baden-Wurttemberg, Germany) in attenuated total reflectance (ATR) mode using a PIKE GladyATR device with diamond ATR unit. The FTIR-ATR spectra were recorded in 4000–400 cm^{-1} range with a resolution of 4 cm^{-1}. The samples were firmly pressed to the surface of the ATR unit by means of a special clamping device.

2.6. Investigation of the Ability of Epoxy Vitrimer Nanocomposite for Welding

The ability of epoxy vitrimer nanocomposites for welding was studied as follows. Rectangular samples ($1.8 \times 4 \times 37$ mm^3) were superimposed on each other. To ensure a good contact between the samples, they were placed in a special clamping device, compressed by ~25%, and left in a drying oven at a temperature of 160 °C for 5–6 h. The overlap area of the rectangular samples was 5×5 mm $= 25$ mm^2. Tensile testing of the welded joints of the samples was carried out on Lloyd LS5, AMETEK STC tensile machine at a speed of 5 mm/min.

3. Results and Discussion

3.1. Mechanical and Thermal Properties of Epoxy/Silica Vitrimer Nanocomposites

The epoxy/silica vitrimer nanocomposites were prepared by mixing 30 wt% colloid dispersion of 10–15 nm spherical silica nanoparticles in MEK with DGEBA and ZAA as the interchain exchange catalyst (Figure 1). The surface silanol groups of the nanoparticles can interact with the epoxy network through H-bonding. It should be noted that no curing catalyst was used during the synthesis. The mechanism of noncatalyzed reaction of aromatic epoxides such as DGEBA with anhydrides is well-studied in the literature [32–35]. It was shown that it starts by the interaction of OH-groups, which are present as impurity in epoxy oligomers, with anhydride giving monoester with a free carboxylic acid group, which, in turn, reacts with an epoxy group forming a diester and a new OH-group that can react with the anhydride (Figure S1, Supplementary Materials).

Conditions of preparation of the epoxy/silica networks are listed in Table 1. The nanocomposites obtained did not dissolve being immersed in trichlorobenzene at 160 °C for 16 h, which confirmed that the polymer is cross-linked, and the exchange reactions occurring in the presence of transesterification catalyst do not induce depolymerization of the network.

Figure 2 shows the FTIR spectra of the V1 and N1-5 samples. In the spectra, one can see the absorption bands of ester groups at 1735 and 1165 cm^{-1} (stretching vibrations of C=O and C–O–C bonds, respectively [36]) arising as a result of the reaction of DGEBA epoxy groups with anhydride MHHPA. At the same time, the characteristic band of the epoxy ring vibration of DGEBA at 916 cm^{-1} (stretching vibrations of C–H bonds of epoxy group) as well as the intense bands of the hardener MHHPA at 1780 and 1858 cm^{-1} (asymmetric ν_{as} and symmetric ν_s stretching vibrations of C=O bond of anhydride group [37]) and at 892 cm^{-1} (stretching vibrations of C–O bond of anhydride group [38]) are absent. These data evidence that the anhydride groups of MHHPA reacted with epoxy groups of DGEBA giving the ester bonds. The appearance of broad absorption between 1000 and 1200 cm^{-1}, culminating at 1100 cm^{-1}, and the signal located at 467 cm^{-1}, corresponding to asymmetric bending vibrations of Si–O–Si bonds [39], proves the incorporation of silica nanoparticles into the epoxy vitrimer matrix.

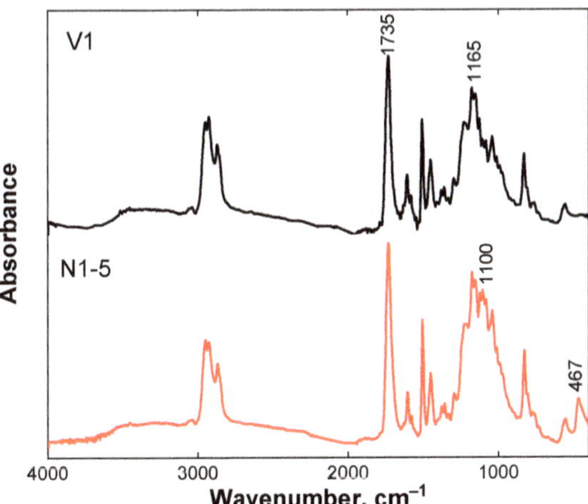

Figure 2. FTIR spectra for epoxy/silica vitrimer nanocomposite N1-5 with 5 wt% silica nanoparticles (red curve) and of the epoxy vitrimer V1 prepared under the same conditions, but without nanoparticles (black curve).

The mechanical properties (elastic modulus, tensile strength, and elongation at break) of the epoxy/silica vitrimer nanocomposites were compared with the properties of the epoxy vitrimers prepared by thermal curing of DGEBA in the presence of MHHPA and ZAA under similar conditions but without silica nanoparticles. It was shown that the introduction of 5–10 wt% of silica nanoparticles in the vitrimer matrix increases the elastic modulus (Table 1, Figure 3). Similar behavior was observed for DGEBA/MHHPA/silica nanocomposites prepared in the absence of transesterification catalyst [24,40–42]. The increase of the modulus at the addition of silica nanoparticles was explained by much higher modulus of silica (G = 29.91 GPa [43], E = 70 GPa [41]) compared to the epoxy matrix and the restriction of polymer chain mobility by nanoparticles. The same reasons seem to be valid for epoxy/silica vitrimer nanocomposites under study.

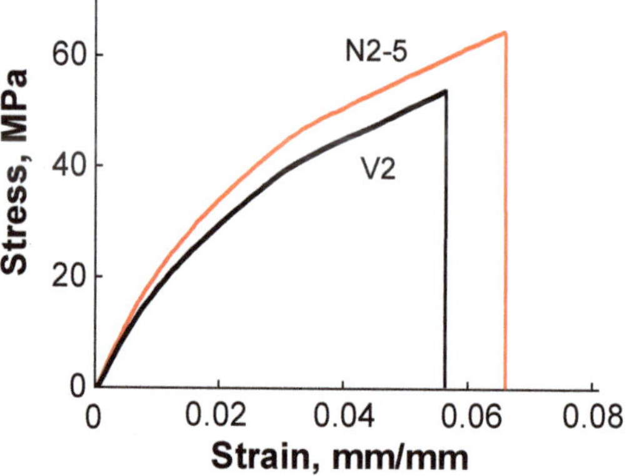

Figure 3. Tensile test for epoxy/silica vitrimer nanocomposite N2-5 with 5wt% silica nanoparticles (red line) and of the epoxy vitrimer V2 prepared under the same conditions, but without nanoparticles (black line).

The empirical Guth equation [44] was used to estimate the Young's modulus E_c of epoxy/silica vitrimers at different content of silica nanoparticles:

$$\frac{E_c}{E_m} = 1 + 2.5V_p + 14.1V_p^2 \tag{2}$$

where V_p is the particle volume fraction, and E_c and E_m are the Young's modulus of the composite and the epoxy matrix, respectively. The results are presented in Table 1. It is seen that the experimental E_c values slightly exceed the calculated ones. Similar behavior was observed for soft epoxy/silica vitrimers [18]. It was attributed to the aggregation of nanoparticles in the epoxy matrix [18].

The increase of the elastic modulus with increasing content of nanoparticles was also found for other vitrimer nanocomposites [19]. Note that to increase the strength of nanocomposites the nanoparticles should interact with the matrix so that the applied stress could be effectively transferred from the matrix to the particles [45]. In the present system, such interactions may be due either to the hydrogen bonding or to transesterification reaction between OH-groups of silica and ester groups of polymer leading to covalent attachment of particles to polymer matrix.

Table 1 and Figure 3 show that the addition of silica nanoparticles increases the tensile stress of vitrimer nanocomposites. Therefore, silica particles make the vitrimer stronger. For instance, in the second series of samples (V2 and N2-5) 5 wt% of silica induces the augmentation of the tensile stress from 54 to 65 MPa (Figure 3), i.e., by 20%. This is close to the results for DGEBA/MHHPA/silica nanocomposites prepared in the absence of transesterification catalyst, for which 5.3 wt% of silica induce the increase of tensile strength by ca. 16% [46]. The observed effect is expected [45] to be mainly due to polymer-particle interactions strengthening the network. At the same time, in soft epoxy vitrimersat room temperature the tensile stress decreases upon addition of silica from 25 to 11 MPa [18]. Therefore, in the present paper we first show that silica particles can strengthen hard epoxy vitrimers by increasing their tensile stress like in the case of ordinary epoxy thermosets, which do not undergo chain exchange reactions.

Figure 3 demonstrates that in the second series of samples (V2 and N2-5) the introduction of silica nanoparticles in the vitrimer matrix increases the elongation at break δ_B from 5.6 to 6.5%. Similar behavior of DGEBA/MHHPA/silica thermoset nanocomposites without transesterification catalyst [42], for which δ_B augments from 2.7 to 4.8% upon addition of 6 wt.% of silica, was attributed to lower crosslinking density of epoxy matrix [42] in the presence of nanoparticles hindering cure reaction. Most probably, the same reason is valid for the vitrimer nanocomposite N2-5 under study. By contrast, in the first series of samples (V1, N1-5 and N1-10, Table 1) synthesized at lower DGEBA concentration the elongation at break decreases from 7.3 to 4.9% in the presence of nanoparticles, similar to the case of soft epoxy/silica vitrimer nanocomposites, where also a two-fold decline of the elongation at break was observed [18]. Such behavior can be related to the attraction between silica particles and polymer matrix. These results suggest that in more dilute system, the nanoparticles do not significantly deteriorate the formation of the epoxy network, and the polymer-particle attraction becomes a dominant factor affecting the elongation at break.

Dynamic mechanical tensile tests at the frequency of 1 Hz were performed. The resulting temperature dependences of storage E' and loss E'' moduli and mechanical loss tangent tan δ are presented in Figure 4. A peak observed on tan δ curve (Figure 4b) determines the α-transition of the cured epoxy network [26]. It indicates that the glass transition temperatures of the nanocomposite and the neat matrix without nanoparticles (samples N2-5 and V2, Table 1) are equal to 150 and 152 °C, respectively. Note that the peak has a small shoulder at temperature around 100 °C, which is more pronounced in nanocomposite as compared to neat epoxy matrix. It may be caused by the fragments of the incompletely formed network as was shown theoretically [47]. Such behavior can result from hindering of the proper curing of polymer caused by nanoparticles. It is interesting that in soft epoxy/silica vitrimer nanocomposites [18] two peaks on the tan δ curves were also observed, but the peak at lower temperature was larger and it was attributed to the α-transition. The smaller peak was not discussed [18].

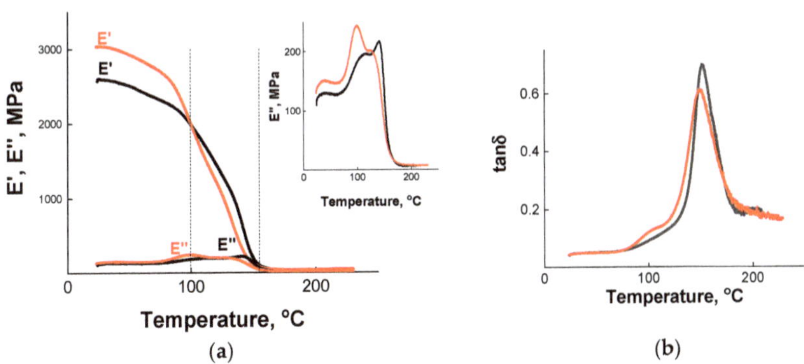

Figure 4. Temperature dependence of storage E' and loss E'' moduli (**a**) and of loss tangent tan δ (**b**) for epoxy/silica vitrimer nanocomposite N2-5 with 5 wt% silica nanoparticles (red curve) and of the epoxy vitrimer V2 prepared under the same conditions, but without nanoparticles (black curve).

Figure 4a shows that E' is much larger than E'' at all studied temperatures, which is expected. At 25 °C, the storage modulus E' of the nanocomposite is equal to 3 GPa, which is appreciably higher than for the neat matrix (2.6 GPa). One can see (Figure 4a) that the storage modulus E' decreases at heating both for nanocomposite and for the unfilled vitrimer, which can be due to the softening of the polymer matrix at increasing temperatures.

In addition, in the case of nanocomposites, some contribution to the lowering of E' comes from the difference between the CTE of the polymer matrix and the silica particles, which induces relaxations in the polymer phase, as was suggested previously [48,49]. Note that below 100 °C, the E' of the nanocomposite is higher than that of the unfilled vitrimer matrix E'. When the temperature rises above 100 °C, the situation changes: the E' of the nanocomposite becomes lower. For instance, at 130 °C the storage modulus equals to 0.82 and 1.16 GPa in the presence and in the absence of nanoparticles, respectively. Finally, in the temperature range from 160 to 250 °C, the difference between both samples vanishes. Therefore, the reinforcing effect of the filler is observed mainly in the glassy state. Probably, when the system approaches T_g, enhanced mobility of polymer chains reduces the adhesion of polymer to the nanoparticles [50]. As a result, the contribution of polymer-particle interactions increasing E' becomes lower than the contribution of polymer defects formed at the network synthesis in the presence of nanoparticles, which decreases E'. Similar behavior was found for epoxy/graphene oxide vitrimer nanocomposites [19]. In this case, below 60 °C the storage modulus E' of nanocomposites was higher than for neat matrix by up to 30%, whereas above this temperature, the situation was reversed, and E' of nanocomposite became up to 6-fold lower than for the neat matrix. Note that in the present system, the decrease of E' of the nanocomposite with respect to neat matrix is much less pronounced.

Higher mobility of polymer chains in the vicinity of T_g induces the rise of the loss modulus E'' (Figure 4a). From Figure 4a (inset) one can see that the temperature dependence of E'' demonstrates two loss maxima: the first one (low-slopped, low-temperature) occurs at 96 and 107 °C, the second one (high-temperature) at 137 and 146 °C for nanocomposite and unfilled vitrimer, respectively. The first maximum at about 100 °C corresponding to the shoulder on the tan δ curve seems to be caused by the fragments of the incompletely formed network as was discussed above. The second (high-temperature) peak of losses associated with a sharp increase of tan δ (Figure 4b) is due to α-transition (glass transition) of the cured epoxy network [46,47]. Note that T_g for the nanocomposite is somewhat lower than for the unfilled network, which may also be due to effective preventing of the complete curing of the epoxy matrix by silica nanoparticles [47]. According to literature data [19], small lowering of T_g can be helpful to achieve a low temperature self-healing of the material. One can suggest that it may be also beneficial for low temperature welding.

TMA was used to compare the thermal properties (T_g, T_v, CTE, and degradation temperature T_d) of the epoxy/silica vitrimer nanocomposites with those of the neat epoxy vitrimer. The data are shown in Figure 5 and Table 2. The first temperature transition (T_g) is associated with devitrification of the material [51]. The second temperature transition (T_v) appears only in the vitrimers [1,3] and reflects defrosting of the topology of polymer networks when the sample is heated. A third transition (T_d) associated with the thermal degradation of the samples is observed at a high temperature [52].

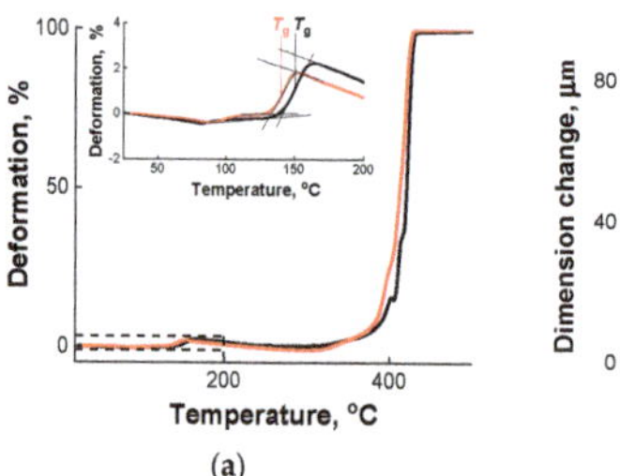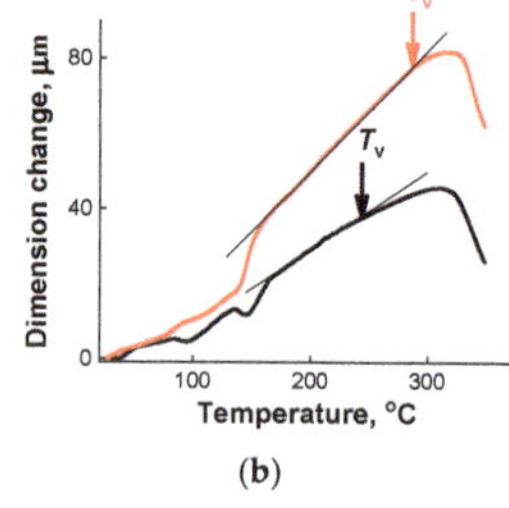

(a)   (b)

Figure 5. Thermomechanical curves (**a**) and expansion as a function of temperature (**b**) for epoxy/silica vitrimer nanocomposite with 5 wt% silica nanoparticles (sample N2-5) (red curve) and for the epoxy vitrimer prepared under the same conditions, but without nanoparticles (sample V2) (black curve).

Table 2. Thermal properties of epoxy/silica vitrimer nanocomposites determined from TMA.

Sample	Glass Transition Temperature T_g, °C	Topology Freezing Temperature T_v, °C	Coefficient of Thermal Expansion CTE × 10^6, μm/(m K)		Temperature of Onset of Thermal Degradation T_d, °C
			Before T_g	Above T_g	
V1	141	230	70	230	396
N1-5	136	280	70	205	396
N1-10	130	290	64	150	395
V2	152	244	70	210	414
N2-5	144	289	70	200	405

The values of T_g were determined from the inflection point of the temperature transition of epoxy/silica thermosets from the glassy to the rubbery state (Figure 5a, inset). Figure 5a and Table 2 show that the addition of SiO_2 nanoparticles slightly diminishes the T_g values, which is consistent with DMA data (Figure 4b). The higher the content of silica the more pronounced is the decrease of T_g (Table 2). Quite similar lowering of T_g values from 144 down to 132 °C upon increasing silica content from 0 to 5.3 wt.% was found previously for DGEBA/MHHPA/silica nanocomposites prepared in the absence of transesterification catalyst [46]. Similar behavior was also observed for epoxy/graphene oxide vitrimer nanocomposites [18]. As stated above, the decrease in T_g may be due to the formation of loose structures [15] when the filler creates steric hindrances for the synthesis of the epoxy network. This contrasts with the behavior observed in soft epoxy/silica nanocomposites, where the T_g slightly increases upon addition of nanoparticles [18]. It indicates that nanoparticles do not interfere the network formation when it is loose. However, when the network is dense (like in the hard thermosets under study), the nanoparticles may hinder the proper cross-linking. This is consistent with the conclusion made above from the analysis of the elongation at break data for soft and hard vitrimer nanocomposites.

The values of T_v temperature were determined according to ref. [1] from a sharp increase in CTE upon heating the sample (Figure 5b), which arises when the exchange reactions make the network more expandable. Our experiments were performed in the penetration mode. In this mode, when the sample is not very soft, its positive deformation is followed by the displacement of the punch, but when the sample becomes very soft (above T_v), the sample expands and the course of deformation during punch penetration becomes negative (Figure 5b). The results of T_v determination are presented in Table 2. It is seen that all samples under study (both in the presence and in the absence of nanoparticles) demonstrate a T_v transition inherent to vitrimers [1,3]. It indicates that the addition of silica does not prevent the transesterication reaction, and the composite should be malleable at high temperature. Note that T_v slightly increases upon incorporation of silica particles. It means that the nanoparticles may stabilize the network topology, and the system needs to be heated more to unfreeze its topological structure. The found T_v values for vitrimers (T_v = 230 and 244 °C for V1 and V2) slightly exceed T_v = 210 °C for a sample obtained by curing DGEBA in the presence of glutaric anhydride and ZAA and determined by dilatometry at heating rate of 5 K/min [3]. The observed small difference in the values of T_v can be due to larger amount of catalyst in the sample reported in [3] (5 mol%), which makes the exchange reactions faster.

A sharp rise in deformation to 100% (Figure 4a) when the sample is heated at high temperatures (in the region about 400 °C) indicates the temperature of the onset of intensive thermal degradation T_d. Table 2 shows that the T_d temperature practically does not change at the introduction of SiO_2 nanoparticles. This is consistent with the literature data [15] for epoxy/carbon fibers nanocomposites.

From Table 2 one can see that the CTE values in the rubbery state (above T_g) are higher than those below T_g since the rubbery material possesses more free volume allowing higher expansion. When the number of added nanoparticles is rather high (10%), they reduce the CTE values especially in the rubbery state (above T_g). Similar behavior was observed for other epoxy/silica nanocomposites [24,43,53]. It was attributed to much lower CTE of the silica nanoparticles as compared to the neat epoxy resin. Additionally, one can expect that interactions of the particles with the epoxy network can restrict the mobility and deformation of the matrix thereby preventing the expansion of the resin matrix upon heating.

Thus, the effect of nanoparticles on the mechanical behavior of epoxy vitrimers at room temperature is quite similar to that observed in the corresponding epoxy thermosets without transesterification catalyst. It concerns, in particular, the increase of the elastic modulus and the tensile strength and the decrease of CTE induced by added nanoparticles. It indicates that nanoparticles make the polymer material stronger and enhance its dimensional stability at heating.

3.2. Investigation of the Ability of Epoxy/Silica Vitrimers for Welding

The welding of epoxy/silica vitrimers was performed by a compress molding. In these experiments, two vitrimer samples ($1.8 \times 4 \times 37$ mm^3) were joined together with an overlap area of 25 mm^2 for few hours under pressure (Figure 6a,b). A welding pressure is necessary because it increases the real contact area on the interface, which is especially important, when the interface is rough. The welding was performed at 160 °C, that is above T_g, but below topology freezing temperature T_v. In the vitrimer, there are some chain exchange reactions even below T_v and we would like to check, whether they can be sufficient to provide a welding in the present system. Taking into account that the temperature of the welding was well below T_v the time of welding was rather long (5–6 h).

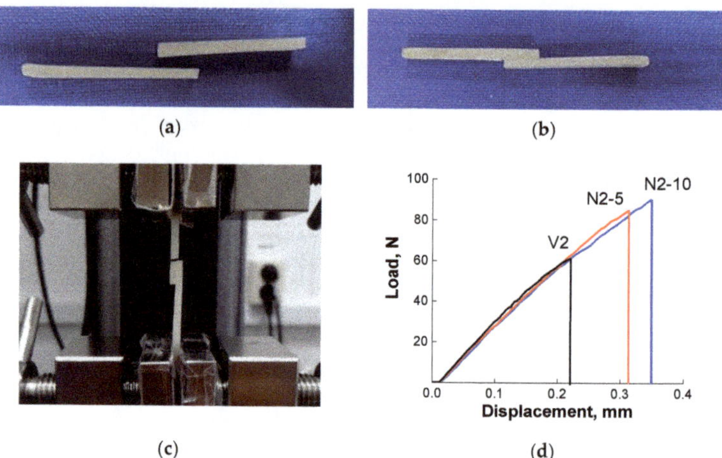

Figure 6. Photographs of the silica/epoxy vitrimer nanocomposite (sample N2-5) before (**a**) and after (**b**) welding at 160 °C for 6 h and after lap-shear test of the welded sample (**c**). Stress–strain curves (**d**) of epoxy/silica vitrimer samples at different content of nanoparticles: 0 (sample V2), 5 (sample N2-5) and 10 wt% (sample N2-10) welded under similar conditions (160 °C, 5 h).

It was shown that this time was enough to weld all the vitrimer samples under study. The adhesion increases with increasing the welding time. For example, samples V2 and N2-5 welded during 5 h disrupt at the welded joint in contrast to the same samples welded during 6 h, which break outside the overlapped part (Table 3). Therefore, the welding effect can be effectively improved by properly increasing the welding time. In further experiments, the welding time was fixed at 6 h.

Table 3. Tensile tests of welded specimens.

Sample	[SiO$_2$], wt%	Welding Time, h	Load, N	Rupture Site
V1	0	6	78	Welding joint
N1-5	5	6	119	Beyond the weld
N1-10	10	6	46	Beyond the weld
V2	0	6	44	Beyond the weld
N2-5	5	6	71	Beyond the weld
V2	0	5	61	Welding joint
N2-5	5	5	85	Welding joint
N2-10	10	5	91	Welding joint

The effect of silica nanoparticles on the ability of epoxy vitrimers for welding and on the strength of the welded joints was studied by tensile test experiments. The results are summarized in Table 3. It is seen that the strength of the welded joint depends on the concentration of silica nanoparticles. For instance, for a first series of samples (V1, N1-5 and N1-10) the welded vitrimer without nanoparticles disrupts in the welding joint at application of 78 N, whereas in the corresponding nanocomposites the rupture occurs outside the welded joint (Table 3). This result indicates to a higher strength of the polymer matrix at the welding site, when nanoparticles are added. Therefore, silica nanoparticles considerably increase the strength of a joint. The positive effect of nanofiller (graphene oxide) on the self-healing properties and shape memory of epoxy vitrimers was discovered by Krishnakumar et al. [19]. Here, we demonstrate similar effect of filler on welding ability. Surprisingly, an increase of the concentration of silica from 5 to 10 wt% diminishes the strength of the joint from 119 to 46 N. It may be due to the presence of some bubbles in this particular sample.

In the second series of samples (V2 and N2-5) prepared at higher monomer concentration both the neat vitrimer and the nanocomposite form rather strong welded joint, which withstands the load (Figure 6c). However, the nanocomposite vitrimer is always stronger (Table 3): it breaks at almost 2-fold higher load than the corresponding neat vitrimer. It means that nanoparticles strengthen the welding joint. Similar conclusion can be drawn from the tensile tests of welded specimens of the third series cured for shorter time (5 h) (Table 3, Figure 6d). Although they are broken at the joint, the breaking load increases with increasing content of nanoparticles. Therefore, the results obtained unambiguously show that nanoparticles significantly increase the strength of the welding joint.

The observed behavior can be explained as follows. According to multiscale modeling of surface welding of thermally induced dynamic covalent network polymers [12], at the beginning of surface welding, an initial chain density gradient drives the polymer chains to diffuse to the interface, where there is a lack of polymer. Then, at the interface the bond exchange reactions between polymer chains belonging to different surfaces occur thereby bridging the two samples together. In the present system, the interchain exchange is the transesterification, which proceeds between ester and hydroxyl groups (Figure 7) producing a new ester and a new hydroxyl group. Increase of the strength of the joint upon addition of nanoparticles indicates that silica particles participate in the welding. They can contribute to the attraction between polymer chains belonging to different welding surfaces by forming hydrogen bonds with them. Moreover, one can suggest that the transesterification reactions can occur not only between epoxy chains, but also on the epoxy/silica interface.

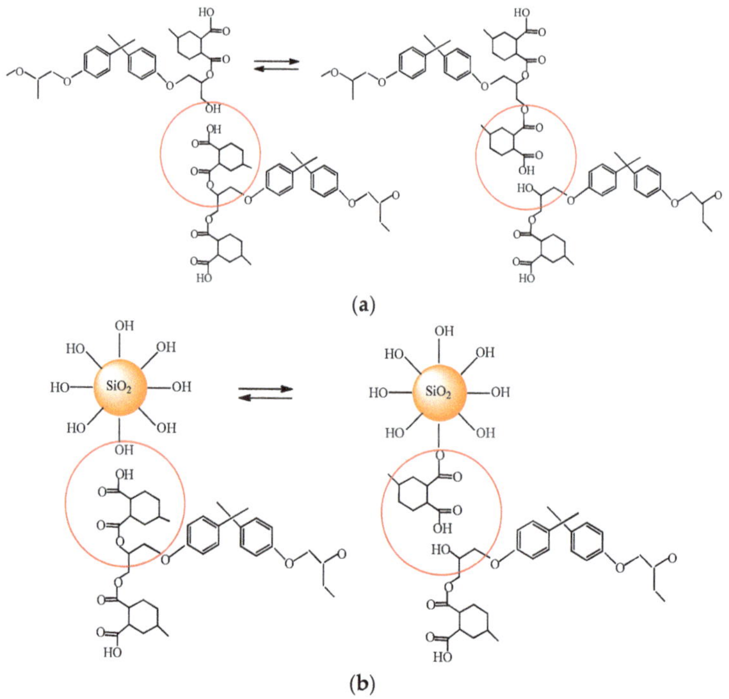

Figure 7. Schematic representation of transesterification reaction between epoxy chains (**a**) and on the epoxy/silica interface (**b**).

Note that in the present study, the welding temperature is well below the T_v. Nevertheless, welding occurs, indicating that some chain exchange reactions proceed even at these conditions, but these reactions are not numerous and therefore the welding requires rather long time. The occurrence of the chain exchange reactions even below T_v can be supported by the fact that epoxy/silica nanocomposite and the corresponding epoxy thermoset (without interchain exchange catalyst) do not weld at 160 °C at all. Therefore, the presence of interchain exchange catalyst is ultimate to weld the samples. The welding of thermoset vitrimer below T_v was previously demonstrated on an example of DGEBA cured with glutaric anhydride in the presence of ZAA [1]. The T_v for this vitrimer was approximately 210 °C [3], and the welding of two samples was performed at 150 °C [1]. In the present paper, for the first time we show that silica nanoparticles promote the welding of vitrimers and make the welding joint stronger.

4. Conclusions

For the first time, hard epoxy/silica thermoset nanocomposites that can change their topology when heated in the presence of interchain exchange catalyst ZAA have been prepared. The topology of the epoxy network is changed as a result of the transesterification reaction via the interaction of carbonyl groups $-C=O$ of ester with hydroxyl groups of products of incomplete curing of the epoxy monomer.

It is shown that in the new nanocomposites, nanoparticles strengthen the vitrimers by increasing their tensile stress from 48 to 60 MPa and elastic modulus from 1.8 to 2.6 GPaand simultaneously improve their dimensional stability at heating, since the CTE in the presence of nanoparticles decreases. This is the first demonstration of the fact that silica particles can reinforce hard epoxy vitrimers by increasing their tensile stress.

By dynamic mechanical analysis it was shown that the reinforcing effect of particles is mainly observed in the glassy state. When the system approaches T_g, enhanced mobility of polymer chains reduces the adhesion of polymer to the nanoparticles thereby diminishing the contribution of polymer-particle interactions to the mechanical properties.

By thermomechanical analysis it was shown that all samples under study (both in the presence and in the absence of nanoparticles) demonstrate a T_v transition inherent to vitrimers indicating that silica particles do not prevent the transesterification reaction. At the same time, nanoparticles slightly increase the topology freezing temperature T_v suggesting that they may stabilize the network topology.

Both by dynamic mechanical and thermomechanical analyses it was evidenced that the addition of SiO_2 slightly diminishes the glass transition temperature of the vitrimers. It was attributed to the formation of loose structures when the incorporated particles hinder the cure reaction. Small decrease of T_g can be beneficial for low temperature welding.

One of the most important findings of this study is the observation that the silica particles enhance the welding of vitrimer samples induced by interchain exchange reactions and make the welding joints stronger. To the best of our knowledge, the last effect of silica particles on the welding was first observed in vitrimers. It opens up wide perspectives for assembling and joining thermoset composites to produce novel repeatedly recyclable vitrimer materials with a combination of desirable properties provided by each of the welded components.

Supplementary Materials: The following are available online at https://www.mdpi.com/article/10.3390/polym13183040/s1, Figure S1: Schematic representation of mechanism of noncatalyzed reaction of aromatic epoxides such as diglycidyl ether of bisphenol A.

Author Contributions: Conceptualization, O.E.P.; methodology, A.I.B. and A.A.A.; investigation, A.I.B., E.S.A. and V.S.M.; resources, A.I.B. and O.E.P.; writing—original draft preparation, A.I.B., V.S.M. and O.E.P.; writing—review and editing, A.I.B. and O.E.P. All authors have read and agreed to the published version of the manuscript.

Funding: This work was supported by the Russian Science Foundation (project No. 17-13-01535).

Institutional Review Board Statement: Not applicable.

Informed Consent Statement: Not applicable.

Data Availability Statement: The data presented in this study is openly available.

Acknowledgments: TMA analyses of thermosets were performed with the financial support from Ministry of Science and Higher Education of the Russian Federation using the equipment of Center for molecular composition studies of A.N. Nesmeyanov Institute of Organoelement Compounds, Russian Academy of Sciences. The authors thank Elena Kharitonova, Alexander Polezhaev, Alexey Kireynov and Boris Lokshin for their help in the study of thermosets using DSC, DMA and FTIR.

Conflicts of Interest: The authors declare no conflict of interest.

References

1. Montarnal, D.; Capelot, M.; Tournilhac, F.; Leibler, L. Silica-like malleable materials from permanent organic networks. *Science* **2011**, *334*, 965–968. [CrossRef] [PubMed]
2. Denissen, W.; Winne, J.M.; Du Prez, F.E. Vitrimers: Permanent organic networks with glass-like fluidity. *Chem. Sci.* **2016**, *7*, 30–38. [CrossRef] [PubMed]
3. Capelot, M.; Montarnal, D.; Tournilhac, F.; Leibler, L. Metal-catalyzed transesterification for healing and assembling of thermosets. *J. Am. Chem. Soc.* **2012**, *134*, 7664–7667. [CrossRef] [PubMed]
4. Ran, Y.; Zheng, L.-J.; Zeng, J.-B. Dynamic crosslinking: An efficient approach to fabricate epoxy vitrimer. *Materials* **2021**, *14*, 919. [CrossRef]
5. Spiesschaert, Y.; Guerre, M.; De Baere, I.; Van Paepegem, W.; Winne, J.M.; Du Prez, F.E. Dynamic curing agents for amine-hardened epoxy vitrimers with short (re)processing times. *Macromolecules* **2020**, *53*, 2485–2495. [CrossRef]
6. Obadia, M.M.; Mudraboyina, B.P.; Serghei, A.; Montarnal, D.; Drockenmuller, E. Reprocessing and recycling of highly cross-linked ion-conducting networks through transalkylation exchanges of C-N bonds. *J. Am. Chem. Soc.* **2015**, *137*, 6078–6083. [CrossRef]
7. Lu, Y.X.; Tournilhac, F.; Leibler, L.; Guan, Z. Making insoluble polymer networks malleable via olefin metathesis. *J. Am. Chem. Soc.* **2012**, *134*, 8424–8427. [CrossRef]
8. Lu, Y.; Guan, Z. Olefin metathesis for effective polymer healing via dynamic exchange of strong carbon − carbon double bonds. *J. Am. Chem. Soc.* **2012**, *134*, 14226–14231. [CrossRef]

9. Imbernon, L.; Oikonomou, E.K.; Norvez, S.; Leibler, L. Chemically crosslinked yet reprocessable epoxidized natural rubber via thermo-activated disulfide rearrangements. *Polym. Chem.* **2015**, *6*, 4271–4278. [CrossRef]

10. Denissen, W.; Rivero, G.; Nicolaÿ, R.; Leibler, L.; Winne, J.M.; Du Prez, F.E. Vinylogous urethane vitrimers. *Adv. Funct. Mater.* **2015**, *25*, 2451–2457. [CrossRef]

11. Yang, Y.; Pei, Z.; Zhang, X.; Tao, L.; Wei, Y.; Ji, Y. Carbon nanotube-vitrimer composite for facile and efficient photo-welding of epoxy. *Chem. Sci.* **2014**, *5*, 3486–3492. [CrossRef]

12. Yu, K.; Shi, Q.; Li, H.; Jabour, J.; Yang, H.; Dunn, M.L.; Wang, T.; Qi, H.J. Interfacial welding of dynamic covalent network polymers. *J. Mech. Phys. Solids.* **2016**, *94*, 1–17. [CrossRef]

13. Liu, H.; Zhang, H.; Wang, H.; Huang, X.; Huang, G.; Wu, J. Weldable, malleable and programmable epoxy vitrimers with high mechanical properties and water insensitivity. *Chem. Eng. J.* **2019**, *368*, 61–790. [CrossRef]

14. Awad, S.A.; Fellows, C.M.; Mahini, S.M. Effects of accelerated weathering on the chemical, mechanical, thermal and morphological properties of an epoxy/multi-walled carbon nanotube composite. *Polym. Test.* **2018**, *66*, 70–77. [CrossRef]

15. Uthaman, A.; Xian, G.; Thomas, S.; Wang, Y.; Zheng, Q.; Liu, X. Durability of an epoxy resin and its carbon fiber-reinforced polymer composite upon immersion in water, acidic, and alkaline solutions. *Polymers* **2020**, *12*, 614. [CrossRef]

16. Kumar, S.; Krishnan, S.; Samal, S.K.; Mohanty, S.; Nayak, S.K. Toughening of petroleum based (DGEBA) epoxy resins with various renewable resources based flexible chains for high performance applications: A review. *Ind. Eng. Chem. Res.* **2018**, *57*, 2711–2726. [CrossRef]

17. Lascano, D.; Quiles-Carrillo, L.; Torres-Giner, S.; Boronat, T.; Montanes, N. Optimization of the curing and post-curing conditions for the manufacturing of partially bio-based epoxy resins with improved toughness. *Polymers* **2019**, *11*, 1354. [CrossRef]

18. Legrand, A.; Soulié-Ziakovic, C. Silica-epoxy vitrimer nanocomposites. *Macromolecules* **2016**, *49*, 5893–5902. [CrossRef]

19. Krishnakumar, B.; Prasanna Sanka, R.V.S.; Binder, W.H.; Park, C.; Jung, J.; Parthasarthy, V.; Rana, S.; Yun, G.J. Catalyst free self-healable vitrimer/graphene oxide nanocomposites. *Compos. Part B Eng.* **2020**, *184*, 107647. [CrossRef]

20. Yue, L.; Ke, K.; Amirkhosravi, M.; Gray, T.G.; Manas-Zloczower, I. Catalyst-free mechanochemical recycling of biobased epoxy with cellulose nanocrystals. *ACS Appl. Bio Mater.* **2021**, *4*, 4176–4183. [CrossRef]

21. Wang, S.; Ma, S.; Cao, L.; Li, Q.; Ji, Q.; Huang, J.; Lu, N.; Xu, X.; Liu, Y.; Zhu, J. Conductive vitrimer nanocomposites enable advanced and recyclable thermo-sensitive materials. *J. Mater. Chem. C* **2020**, *8*, 11681–11686. [CrossRef]

22. Liu, Y.-Y.; Liu, G.-L.; Li, Y.-D.; Weng, Y.; Zeng, J.-B. Biobased high-performance epoxy vitrimer with UV shielding for recyclable carbon fiber reinforced composites. *ACS Sustain. Chem. Eng.* **2021**, *9*, 4638–4647. [CrossRef]

23. Kaiser, T. Highly crosslinked polymers. *Prog. Polym. Sci.* **1989**, *14*, 373–450. [CrossRef]

24. Sprenger, S. Epoxy resin composites with surface-modified silicon dioxide nanoparticles: A review. *J. Appl. Polym. Sci.* **2013**, *130*, 1421–1428. [CrossRef]

25. Pourrajabian, A.; Dehghan, M.; Javed, A.; Wood, D. Choosing an appropriate timber for a small wind turbine blade: A comparative study. *Renew. Sustain. Energy Rev.* **2019**, *100*, 1–8. [CrossRef]

26. Ruiz, Q.; Pourchet, S.; Placet, V.; Plasseraud, L.; Boni, G. New eco-friendly synthesized thermosets from isoeugenol-based epoxy resins. *Polymers* **2020**, *12*, 229. [CrossRef] [PubMed]

27. Yuasa, S.; Okabayashi, M.; Ohno, H.; Suzuki, K.; Kusumoto, K. Amorphous, Spherical Inorganic Compound and Process for Preparation Thereof. U.S. Patent 4,764,497A, 16 August 1988.

28. Barabanova, A.I.; Pryakhina, T.A.; Afanas'ev, E.S.; Zavin, B.G.; Vygodskii, Y.S.; Askadskii, A.A.; Philippova, O.E.; Khokhlov, A.R. Anhydride modified silica nanoparticles: Preparation and characterization. *Appl. Surf. Sci.* **2012**, *258*, 3168–3172. [CrossRef]

29. Barabanova, A.I.; Afanas'ev, E.S.; Askadskii, A.A.; Khokhlov, A.R.; Philippova, O.E. Synthesis and propertiesof epoxy networks with a tunable matrix. *Polym. Sci. Ser. A* **2019**, *61*, 375–381. [CrossRef]

30. Turi, E.A. *Thermal Characterization of Polymeric Materials*; Academic Press: New York, NY, USA, 1981.

31. He, Y. Thermal characterization of overmolded underfill materials for stacked chip scale packages. *Thermochim. Acta* **2005**, *433*, 98–104. [CrossRef]

32. Trappe, V.; Burchard, W.; Steinmann, B. Anhydride-cured epoxies via chain reaction. 1. The phenyl glycidyl ether/phthalic acid anhydride system. *Macromolecules* **1991**, *24*, 4738–4744. [CrossRef]

33. Fisch, W.; Hofmann, W. The curing mechanism of epoxy resins. *J. Appl. Chem.* **1956**, *6*, 429–441. [CrossRef]

34. Fisch, W.; Hofmann, W. Chemischeraufbau von gehärtetenepoxyharze. III. Mitteilungüberchemie der epoxyharze. *Makromol. Chem.* **1961**, *44*, 8–23. [CrossRef]

35. Fischer, R.F. Polyesters from epoxides and anhydrides. *Ind. Eng. Chem.* **1960**, *52*, 321–323. [CrossRef]

36. Spegazzini, N.; Ruisánchez, I.; Larrechi, M.S. MCR-ALS for sequential estimation of FTIR-ATR spectra to resolve a curing process using global phase angle convergence criterion. *Anal. Chim. Acta* **2009**, *642*, 155–162. [CrossRef]

37. Rohde, B.J.; Robertson, M.L.; Krishnamoorti, R. Concurrent curing kinetics of an anhydride-cured epoxy resin and polydicyclopentadiene. *Polymer* **2015**, *69*, 204–214. [CrossRef]

38. Tao, S.G.; Chow, W.S. Thermal properties, curing characteristics and water absorption of soybean oil-based thermoset. *Express Polym. Lett.* **2011**, *5*, 480–492. [CrossRef]

39. Kiselev, A.V.; Lygin, V.I. *Infrared Spectra of Surface Compounds*; Nauka: Moscow, Russia, 1972.

40. Johnsen, B.B.; Kinloch, A.J.; Mohammed, R.D.; Taylor, A.C.; Sprenger, S. Toughening mechanisms of nanoparticle-modified epoxy polymers. *Polymer* **2007**, *48*, 530–541. [CrossRef]

41. Hsieh, T.H.; Kinloch, A.J.; Masania, K.; Taylor, A.C.; Sprenger, S. The mechanisms and mechanics of the toughening of epoxy polymers modified with silica nanoparticles. *Polymer* **2010**, *51*, 6284–6294. [CrossRef]
42. Tang, L.C.; Zhang, H.; Sprenger, S.; Ye, L.; Zhang, Z. Fracture mechanisms of epoxy-based ternary composites filled with rigid-soft particles. *Compos. Sci. Technol.* **2012**, *72*, 558–565. [CrossRef]
43. Dittanet, P.; Pearson, R.A. Effect of silica nanoparticle size on toughening mechanisms of filled epoxy. *Polymer* **2012**, *53*, 1890–1905. [CrossRef]
44. Guth, E. Theory of filler reinforcement. *J. Appl. Phys.* **1945**, *16*, 20–25. [CrossRef]
45. Fu, S.-Y.; Feng, X.-Q.; Lauke, B.; Mai, Y.-W. Effects of particle size, particle/matrix interface adhesion and particle loading on mechanical properties of particulate–polymer composites. *Composites Part B* **2008**, *39*, 933–961. [CrossRef]
46. Liu, S.; Zhang, H.; Zhang, Z.; Sprenger, S. Epoxy resin filled with high volume content nano-SiO_2 particles. *J. Nanosci. Nanotechnol.* **2009**, *9*, 1412–1417. [CrossRef] [PubMed]
47. Askadskii, A.A.; Barabanova, A.I.; Afanasev, E.S.; Kagramanov, N.D.; Mysova, N.E.; Ikonnikov, N.S.; Kharitonova, E.P.; Lokshin, B.V.; Khokhlov, A.R.; Philippova, O.E. Revealing defects hampering the formation of epoxy networks with extremely high thermal properties: Theory and experiments. *Polym. Test.* **2020**, *90*, 106645. [CrossRef]
48. Preghenella, M.; Pegoretti, A.; Migliaresi, C. Thermo-mechanical characterization of fumed silica-epoxy nanocomposites. *Polymer* **2005**, *46*, 12065–12072. [CrossRef]
49. Tarrío-Saavedra, J.; López-Beceiro, J.; Naya, S.; Gracia, C.; Artiaga, R. Controversial effects of fumed silica on the curing and thermomechanical properties of epoxy composites. *Express Polym. Lett.* **2010**, *4*, 382–395. [CrossRef]
50. Cui, W.; You, W.; Sun, Z.; Yu, W. Decoupled polymer dynamics in weakly attractive poly(methyl methacrylate)/silica nanocomposites. *Macromolecules* **2021**, *54*, 5484–5497. [CrossRef]
51. Barabanova, A.I.; Lokshin, B.V.; Kharitonova, E.P.; Afanas'ev, E.S.; Askadskii, A.A.; Philippova, O.E. Curing cycloaliphatic epoxy resin with 4-methylhexahydrophthalic anhydride: Catalyzed vs. uncatalyzed reaction. *Polymer* **2019**, *178*, 121590. [CrossRef]
52. Askadskii, A.A. *Computational Materials Science of Polymers*; International Science Publishing: Cambridge, UK, 2003.
53. Pethrick, R.A.; Miller, C.; Rhoney, I. Influence of nanosilica particles on the cure and physical properties of an epoxy thermoset resin. *Polym. Int.* **2010**, *59*, 236–241. [CrossRef]

MDPI

Article

The Influence of Bi_2O_3 Nanoparticle Content on the γ-ray Interaction Parameters of Silicon Rubber

Mahmoud I. Abbas [1], Ahmed M. El-Khatib [1], Mirvat Fawzi Dib [1], Hoda Ezzelddin Mustafa [2], M. I. Sayyed [3,4] and Mohamed Elsafi [1,*]

[1] Physics Department, Faculty of Science, Alexandria University, Alexandria 21511, Egypt; mabbas@physicist.net (M.I.A.); elkhatib60@yahoo.com (A.M.E.-K.); mirvatdib2018@gmail.com (M.F.D.)
[2] Khalifa Medical Center, Abu Dhabi W13-01, United Arab Emirates; hezzeddin68m@gmail.com
[3] Department of Physics, Faculty of Science, Isra University, Amman 11622, Jordan; dr.mabualssayed@gmail.com
[4] Department of Nuclear Medicine Research, Institute for Research and Medical Consultations, Imam Abdulrahman bin Faisal University, Dammam 31441, Saudi Arabia
* Correspondence: mohamedelsafi68@gmail.com

Abstract: In this study, synthetic silicone rubber (SR) and Bi_2O_3 micro- and nanoparticles were purchased. The percentages for both sizes of Bi_2O_3 were 10, 20 and 30 wt% as fillers. The morphological, mechanical and shielding properties were determined for all the prepared samples. The Linear Attenuation Coefficient (LAC) values of the silicon rubber (SR) without Bi_2O_3 and with 5, 10, 30 and 30% Bi_2O_3 (in micro and nano sizes) were experimentally measured using different radioactive point sources in the energy range varying from 0.06 to 1.333 MeV. Additionally, we theoretically calculated the LAC for SR with micro-Bi_2O_3 using XCOM software. A good agreement was noticed between the two methods. The NaI (Tl) scintillation detector and four radioactive point sources (Am-241, Ba-133, Cs-137 and Co-60) were used in the measurements. Other shielding parameters were calculated for the prepared samples, such as the Half Value Layer (HVL), Mean Free Path (MFP) and Radiation Protection Efficiency (RPE), all of which proved that adding nano-Bi_2O_3 ratios of SR produces higher shielding efficiency than its micro counterpart.

Keywords: silicon rubber; nano-Bi_2O_3; LAC; RPE; HVL

Citation: Abbas, M.I.; El-Khatib, A.M.; Dib, M.F.; Mustafa, H.E.; Sayyed, M.I.; Elsafi, M. The Influence of Bi_2O_3 Nanoparticle Content on the γ-ray Interaction Parameters of Silicon Rubber. *Polymers* **2022**, *14*, 1048. https://doi.org/10.3390/polym14051048

Academic Editors: Jesús-María García-Martínez and Emilia P. Collar

Received: 7 February 2022
Accepted: 3 March 2022
Published: 6 March 2022

Publisher's Note: MDPI stays neutral with regard to jurisdictional claims in published maps and institutional affiliations.

1. Introduction

In medical facilities, such as hospitals, clinics, outpatient care centers, radiological centers and dental facilities, where ionizing radiation is widely utilized, planning is compulsory to protect patients and medical staff who are usually exposed to different types of radiation. For this reason, it is important to use radiation protection materials, whether or not these materials are worn, such as eyeglasses, neck guards or an apron [1–4]. Moreover, it is important to utilize specific materials to insulate the walls of the medical facilities in order to prevent radiation leakage into the surrounding environment. This applies not only to the medical facilities, but also all facilities that utilize gamma radiation or X-ray, such as universities and research laboratories, nuclear power plants, and factories [5,6].

The attenuation properties for the radiation protection medium must be accurately known when planning to design any facility that uses gamma rays and X-rays, so that appropriate protection is provided for patients, workers, visitors, and the surrounding environment [7–9]. The radiation protection properties of a medium depend on its density as well as the chemical composition of the medium's constituent materials. Thickness is also considered as another factor that affects the shielding properties of a given medium. The traditional materials that are practically used in radiation protection applications have several drawbacks.

Some of these materials are expensive and some are heavy, and this limits their use in practical applications. For example, tungsten has a higher number of attenuation factors than lead; it also has a high cost and this prevents tungsten from being widely used in real applications. Recently, researchers turned to the production of radiation protective clothing that is characterized by its low cost, light weight, easy use, comfort and, most importantly, its protection of workers in the field of ionizing radiation [10–12]. In this regard, some additives are usually added to the flexible materials to fabricate protective garments, such as aprons or curtains. In practical applications, vinyl, polymers and rubbers are one of the most widespread materials used as matrix materials to obtain flexible protection materials, while bismuth, tungsten and antimony powders are used as additives. The importance of such additives is to increase the possibilities of the photon interactions with the atoms of the prepared flexible protective materials.

As is known, the polymers and plastics have a relatively low density and this, to a certain degree, produces their ability to attenuate photons. Therefore, these plastic materials are usually used to attenuate low-energy radiation [13–15]. In order to increase its density and thus improve its shielding performance, especially if it is exposed to photons of medium energy, bismuth is used [16,17]. Silicone rubber (SR) is an important matrix material that has good elasticity features. In the past few years, some researchers used silicone rubber to develop flexible radiation shielding materials. Kameesy et al. [18] fabricated SR sheets filled with four concentrations of PbO. They experimentally evaluated the radiation attenuation factors for the prepared SR-PbO campsites using different radioactive sources. They found that adding PbO to the SR enhances the physico-mechanical features. Gong et al. [19] fabricated a novel radiation protection composite based on methyl vinyl silicone rubber. The authors found that when benzophenone is added to the matrix, a notable enhancement in the radiation resistance occurs. Based on their results, the transmission of the photon with energy of 0.662 MeV through a sample thickness of 2 cm is only 0.7. Özdemir and Yılmaz [20] prepared a mixed radiation shielding via 3-layered polydimethylsiloxane rubber composite. The three layers were composed of hexagonal boron nitride, B_2O_3 and Bi_2O_3. They developed a shielding material that possesses a lead equivalent thickness of 0.35 mm Pb. Chai et al. [21] prepared new flexible shielding material using methyl vinyl silicone rubber. They used zinc borate, B_4C and hollow beads as filler materials. They evaluated the neutron shielding performance of their flexible material of the thermal neutron transmission technique with the help of an Am–Be radiation source. However, even though different research groups studied the SR as flexible shielding materials, there is still an urgent need for the further development of novel flexible materials using Bi_2O_3 as a filler in nano and micro sizes. Therefore, in this study, we develop a new flexible material against X-ray and γ-ray photons based on Bi_2O_3 nanoparticle content.

2. Materials and Methods

2.1. Matrix

Vulcanized silicone rubber was used as a flexible matrix material. The most common form of silicone is the polydimethylsiloxane polymer, which is liquid in origin. This polymer is a rigid structure of elastomers transformed by catalyzed cross-linking reactions [22]. To obtain catalyzed cross-linking reactions, a stiffener with 4% (by weight) must be added to the silicon rubber. The specific gravity of SR was 1.12 g/mL and the elongation was 350%. The main elements of SR are hydrogen, carbon, oxygen and silicon, as shown in Figure 1.

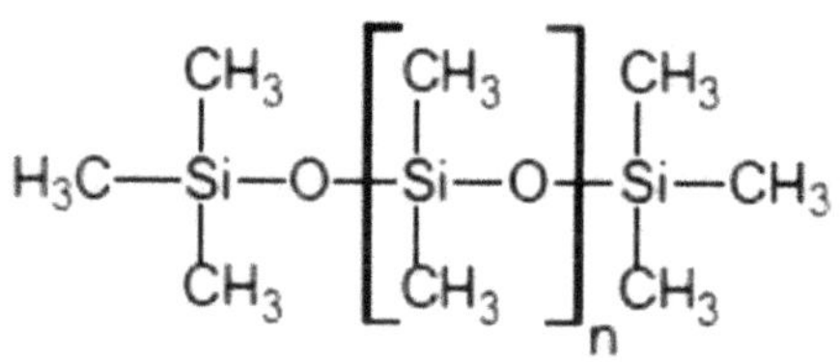

Figure 1. The structure of silicon rubber.

2.2. Fillers

Bismuth oxide (Bi_2O_3) of micro and nano sizes was used as a filler in the composite. Before adding it to the solution, a transmission electron microscope (TEM) is used to image the powder to ensure the size of the particles, as shown in Figure 2. The average size of the microparticles was 15 ± 5 µm, while the average size of the nanoparticles was 30 ± 5 nm.

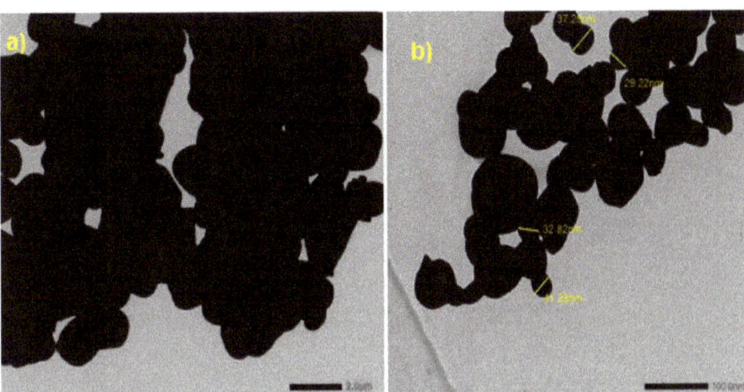

Figure 2. TEM images for (**a**) Bi_2O_3 microparticles and (**b**) Bi_2O_3 nanoparticles.

2.3. Composites

Seven different SR samples were prepared. Codes, compositions and densities of the prepared samples are tabulated in Table 1. The homogenous mixtures (liquid SR + micro- or nanoparticles of Bi_2O_3 + stiffener) were poured into cylinder molds, which had a 3 cm diameter and different thicknesses (0.3, 0.66, 0.93 and 1.3 cm). The prepared samples waited for 24 h to become elastic-solid materials. The density of the SR sample is measured via the mass/volume, the volume of the sample is calculated by ($\pi r^2 \cdot x$), where r is the radius and x is the thickness of the measured sample.

Table 1. Codes, compositions and densities of the prepared SR samples.

Code	Compositions (wt%)				Density (g/cm³)
	SR	**Micro-Bi_2O_3**	**Nano-Bi_2O_3**	**Stiffener**	
SR-0	100	-	-		1.191
SR-5m	95	5	-		1.301
SR-5n	95	-	5		1.351
SR-10m	90	10	-	4	1.368
SR-20m	80	20	-		1.509
SR-30m	70	30	-		1.684
SR-30n	70	-	30		1.713

2.4. Morphological Images

A scanning electron microscope (SEM) of JSM-5300, JEOL model, Tokyo, Japan was used for scanning the images of the prepared SR samples. The samples were coated using an ion sputtering coating device (JEOL-JFC-1100E, Tokyo, Japan), and then the samples were placed inside the electron microscope with an operating voltage of 20 keV [23].

2.5. Mechanical Properties

The tensile strength, Young's modulus and elongation at break were determined for the present SR samples using an electronic tensile testing machine (model 1425, Germany), according to standard methods with ASTM D412. The Shore hardness was measured according to ASTM D2240.

2.6. Shielding Properties

The linear attenuation coefficient (*LAC*) was measured for all discussed SR samples using the narrow beam technique of gamma ray spectroscopy in a radiation physics laboratory (Faculty of Science, Alexandria University, Alexandria, Egypt). The devices used in this method were the detector, collimator and radioactive point sources. An NaI (Tl) cylindrical scintillation detector with a ($3'' \times 3''$) dimension, a relative efficiency of 15% and an energy value of Cs-137 (0.662 MeV) was used. The inner diameter of the lead collimator was 8mm and the outer diameter was 100 mm. The point radioactive sources were chosen to cover a wide range of energy, where four radioactive sources were used as follows: Am-241 (0.06 MeV), Ba-133 (0.081, 0.356 MeV), Cs-137 (0.662 MeV) and Co-60 (1.173, 1.333 MeV) [24–29]. The illustration of the experimental setup is shown in Figure 3.

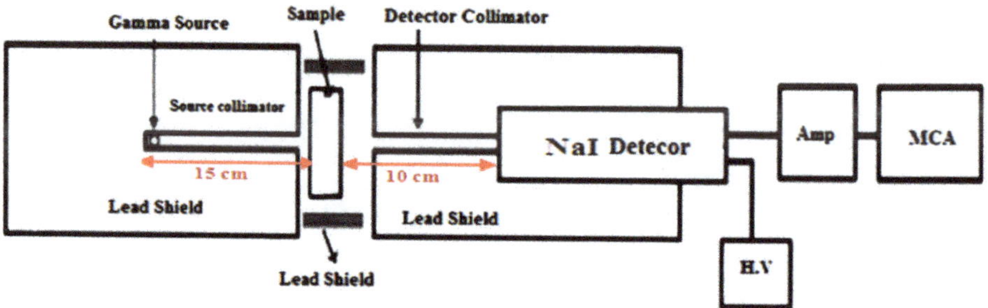

Figure 3. The experimental setup used to measure the attenuation coefficient.

The intensity (I_0), count rate (N_0) or area under the peak (A_0) were measured for all energies in a case without the SR sample, and then the sample was placed between the source and detector and the count rate (N) or area under the peak (A) was measured at the same time. The *LAC* was measured experimentally using the following equation [30,31]:

$$LAC = \frac{1}{x} \, ln\frac{N_0}{N} = \frac{1}{x} \, ln\frac{A_0}{A} \tag{1}$$

The experimental values of the *LAC* for SR and the micro filler were compared to the results obtained from the *XCOM* software [32,33]. The relative deviation between the two results is calculated by the following:

$$Dev(\%) = \frac{LAC_{xcom} - LAC_{exp}}{LAC_{exp}} \times 100 \tag{2}$$

While the relative increase between the results of the *LAC* of the micro and nano fillers are evaluated via the following:

$$R \cdot I(\%) = \frac{LAC_{nano} - LAC_{micro}}{LAC_{micro}} \times 100 \tag{3}$$

The other radiation attenuation parameters are based on the *LAC*, such as the Half Value Layer (*HVL*), which represents the thickness needed to reduce the initial intensity to its half value; the Mean Free Path (*MFP*), which represents the path of the photon inside the sample without any interactions; and the Tenth Value Layer (*TVL*), which represents the thickness needed to reduce the initial intensity to its tenth value and can be estimated from the following equation [34–36]:

$$HVL = \frac{\ln 2}{LAC}, \quad MFP = \frac{1}{LAC}, \quad TVL = \frac{\ln 10}{LAC} \tag{4}$$

The efficiency of shielding materials is estimated by an important parameter called the Radiation Protection Efficiency (*RPE*) and calculated using the following equation [37–39]:

$$RPE(\%) = [1 - \frac{N}{N_0}] \times 100 \tag{5}$$

3. Results and Discussion

3.1. SEM Results

SEM images of the prepared samples were scanned and showed, in general, a good distribution of Bi_2O_3 with the SR composite. On the other hand, the distribution of nanoparticles was better than the microparticles, as shown in Figure 4, which means that the SR containing nanoparticles of Bi_2O_3 has a higher surface area and lower porosity, compared to the same contents of SR containing microparticles of Bi_2O_3. as Additionally, the porosity of SR containing nanoparticles of Bi_2O_3 is low, which led to an increase in the mechanical and shielding properties.

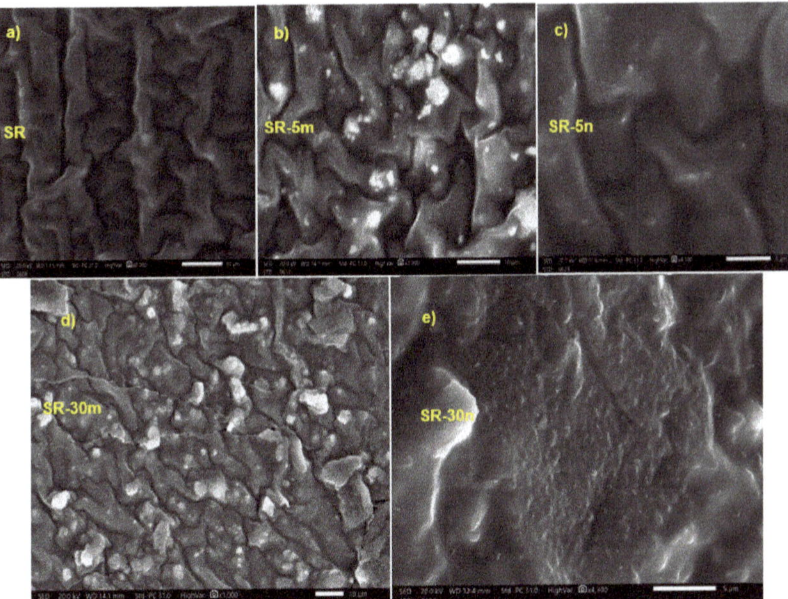

Figure 4. SEM images of the prepared samples of (**a**) SR, (**b**) SR-5m, (**c**) SR-5n, (**d**) SR-30m and (**e**) SR-30n.

3.2. Mechanical Results

In the case of free SR, the mechanical properties (MPs) of SR composites are relatively poor. The MPs of the SR/micro- and nano Bi_2O_3 composites are plotted in Figure 5A–D. These figures show the variability of tensile strength, Young's modulus and elongation at break with different concentrations of micro- and nano-Bi_2O_3 as fillers. The results show that increasing the filler load leads to a significant increase in the tensile strength, Young's modulus and elongation at break of up to 30 wt%.

The results also show that the addition of nano-Bi_2O_3 produces a greater an increase in tensile strength, Young's modulus and elongation at break than micro-Bi_2O_3 with the same percentage. The concentration of the filler was increased to 40%, and the same mechanical properties were studied as before, and it was found that it was less than 30%, and this is what made us conduct a comprehensive study with a maximum of 30% for micro- and nano-Bi_2O_3 as a filler in the SR. Low mechanical properties at 40 wt% of the filler content is likely due to the accumulation of filler material in different rubber layers. The hardness was increased with the increase in the filler contents, and this was normal because the addition of filler in the SR leads to an increase in material hardness.

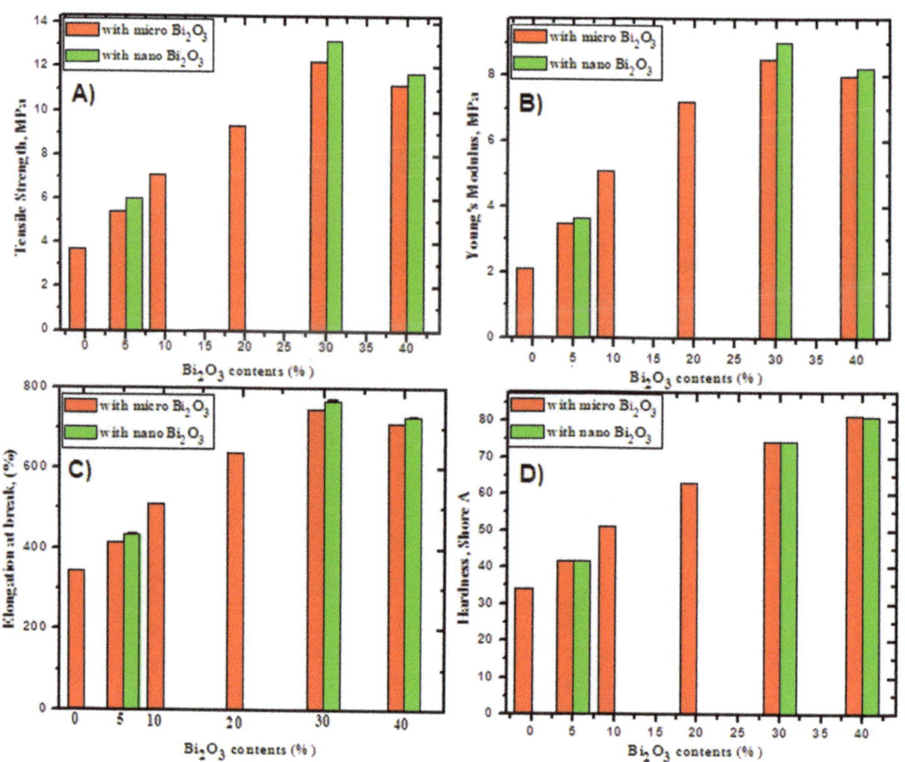

Figure 5. The mechanical properties of SR with different contents of micro- and nano-Bi_2O_3. (**A**) Tensile Strength, MPa, (**B**) Young modulus, MPa, (**C**) Elongations (%), (**D**) Hardness, Shore A.

3.3. Shielding Results

In order to obtain the linear attenuation coefficient experimentally, we represented the relation between Ln (I/I_0) and the thickness of the samples, according to the Lambert–Beer law. The slope of the straight line is the absolute value of the *LAC*. We represent the reduction in the intensity of the photons as a function of the thickness for four samples in Figure 6a–d. In this figure, we show the results for the following samples: SR-5m, SR-5n, SR-30m and SR-30n. The other samples have the same trend shown in this figure, so we did not present the data for the remaining samples in Figure 6a–d.

As one can see in this figure, the slope is negative, which implies that the transmission of the photons through the prepared silicone rubber samples decrease with increasing the thickness from 3 to 13 mm. The slope of the rubber silicon sample, SR-5m, is -0.1502 at 0.356 MeV, and, as can be seen from Table 2, the *LAC* for this sample at 0.356 MeV is 0.1502 cm^{-1}. The *LAC* values for all the prepared rubber silicon with different amounts of micro- and nano-Bi_2O_3 is summarized in Table 3.

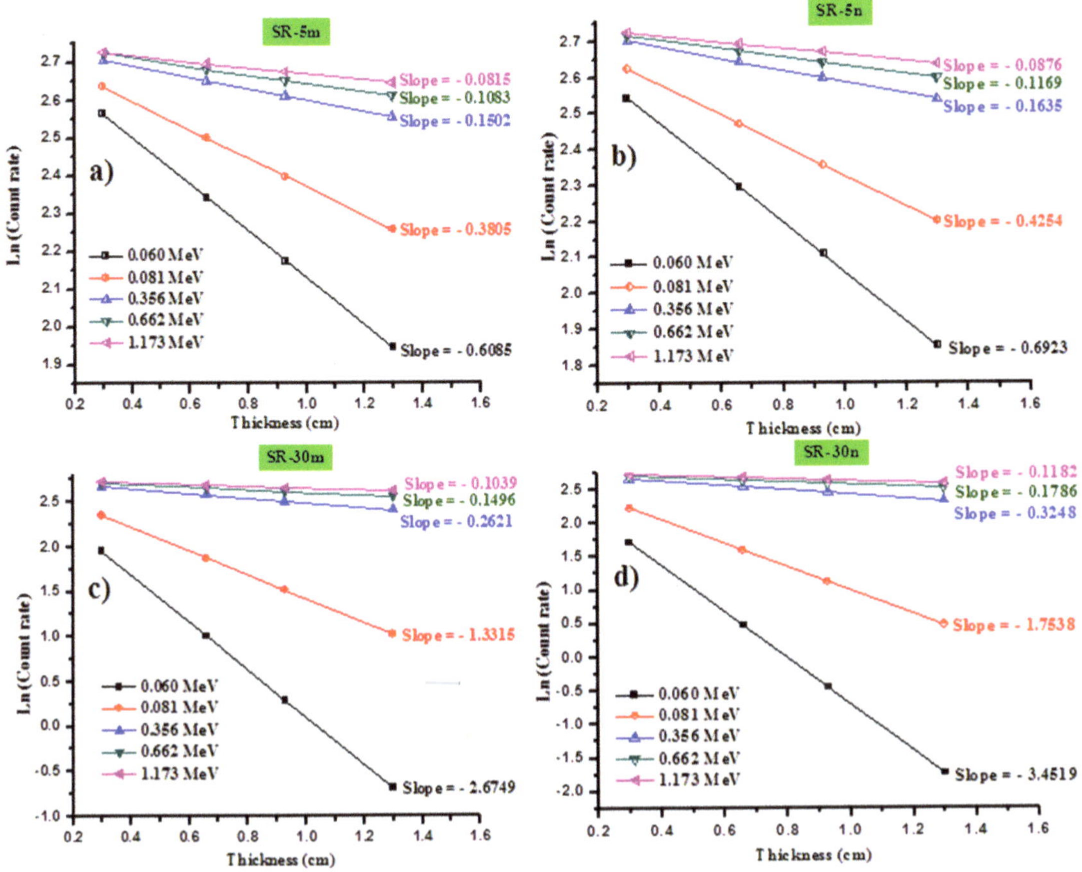

Figure 6. Graphical representation of the reduction in the intensity of the photons as a function of the thickness for (**a**) SR-5m, (**b**) SR-5n, (**c**) SR-30m and (**d**) SR-30n.

Table 2. Linear attenuation coefficient of silicon rubber with different additives (fraction by weight).

Energy (MeV)	SR-0			SR-10m			SR-20m		
	XCOM	EXP	Dev (%)	XCOM	EXP	Dev (%)	XCOM	EXP	Dev (%)
0.060	0.3097	0.3059	1.25	0.9620	0.9464	1.65	1.7499	1.7215	1.65
0.081	0.2456	0.2384	3.01	0.544	0.5305	2.52	0.9042	0.8860	2.05
0.356	0.1354	0.1351	0.25	0.171	0.1657	3.25	0.2141	0.2113	1.35
0.662	0.1043	0.1033	0.98	0.118	0.1154	1.85	0.1336	0.1308	2.14
1.173	0.0794	0.0779	1.89	0.087	0.0861	0.62	0.0954	0.0952	0.28
1.333	0.0744	0.0728	2.11	0.081	0.0796	1.63	0.0889	0.0876	1.48

Table 3. Linear attenuation coefficient of bulk and nano samples.

Energy (MeV)	SR-5					SR-30				
	XCOM	SR-5m	Dev (%)	SR-5n	R.I (%)	XCOM	SR-30m	Dev (%)	SR-30m	R.I (%)
0.060	0.6213	0.6085	2.11	0.6923	12.11	2.7206	2.6749	1.71	3.4519	22.51
0.081	0.3881	0.3805	1.99	0.4254	10.55	1.3481	1.3315	1.25	1.7538	20.08
0.356	0.1525	0.1502	1.52	0.1635	8.14	0.2672	0.2621	1.95	0.3248	15.32
0.662	0.1106	0.1083	2.15	0.1169	7.31	0.1534	0.1496	2.54	0.1786	13.22
1.173	0.0829	0.0815	1.62	0.0876	6.88	0.1063	0.1039	2.31	0.1182	11.46
1.333	0.0775	0.0756	2.55	0.0810	6.67	0.0987	0.0976	1.22	0.1098	11.11

In this study, the linear attenuation coefficient (LAC) values of the SR without Bi_2O_3 and with 5, 10, 20 and 30% Bi_2O_3 (in micro and nano sizes) were experimentally measured using different radioactive point sources in the energy range varying from 0.06 to 1.333 MeV. Additionally, we theoretically calculated the LAC for the SR with micro-Bi_2O_3 using XCOM software. The other parameters based on LAC were calculated, such as HVL, MFP and TVL, and tabulated in Table 4.

Table 4. The HVL, MFP and TVL for all the micro and nano prepared samples.

Shielding Parameters	Energy (MeV)	SR-0	SR-5m	SR-5n	SR-10m	SR-20m	SR-30m	SR-30n
HVL (cm)	0.060	2.2380	1.1156	1.0012	0.7206	0.3961	0.2548	0.2008
	0.081	2.8221	1.7860	1.6294	1.2745	0.7666	0.5142	0.3952
	0.356	5.1177	4.5465	4.2399	4.0525	3.2373	2.5943	2.1339
	0.662	6.6456	6.2649	5.9318	5.8966	5.1878	4.5185	3.8818
	1.173	8.7318	8.3659	7.9165	8.0008	7.2628	6.5214	5.8647
	1.333	9.3223	8.9437	8.5600	8.5649	7.7959	7.0193	6.3156
MFP (cm)	0.060	3.2287	1.6095	1.4444	1.0395	0.5715	0.3676	0.2897
	0.081	4.0714	2.5767	2.3507	1.8387	1.1060	0.7418	0.5702
	0.356	7.3833	6.5592	6.1169	5.8466	4.6705	3.7428	3.0786
	0.662	9.5876	9.0383	8.5577	8.5069	7.4844	6.5188	5.6002
	1.173	12.5973	12.0695	11.4212	11.5427	10.4779	9.4084	8.4610
	1.333	13.4492	12.9030	12.3495	12.3565	11.2472	10.1267	9.1114
TVL (cm)	0.060	7.4345	3.7060	3.3259	2.3936	1.3158	0.8463	0.6670
	0.081	9.3748	5.9330	5.4126	4.2338	2.5466	1.7080	1.3129
	0.356	17.0007	15.1031	14.0846	13.4623	10.7542	8.6181	7.0887
	0.662	22.0762	20.8115	19.7049	19.5879	17.2335	15.0102	12.8949
	1.173	29.0063	27.7910	26.2982	26.5780	24.1264	21.6636	19.4822
	1.333	30.9680	29.7103	28.4357	28.4519	25.8976	23.3176	20.9799

The comparison between the experimental and theoretical LAC for the SR (free Bi_2O_3) and SR with 20% micro-Bi_2O_3 is plotted in Figure 7. We notice good comparability between both LAC values measured in the lab and those calculated by XCOM. This is true for most tested energies, however, we found some minor differences between both approaches, and this is acceptable since usually anyone can find some small errors in the experimental part, but generally the experimental results match the XCOM results in an acceptable manner. This is an essential and important step since it provides confidence in the accuracy of the geometry utilized in the lab for the determination of LAC for the SR and SR/micro-Bi_2O_3 samples. We also calculated the error (Dev.%) between the experimental and XCOM data.

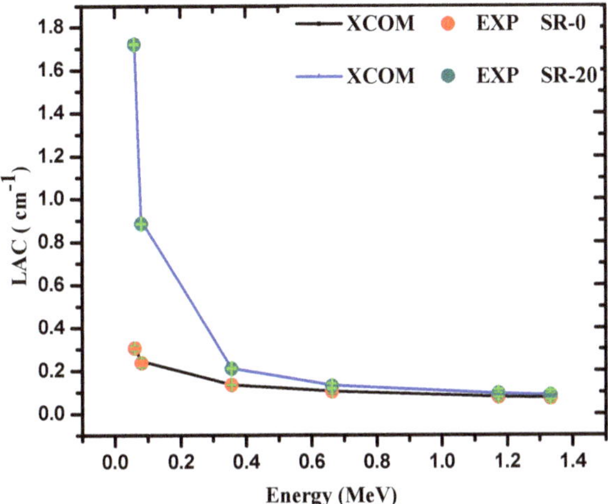

Figure 7. Comparison between the experimental and XCOM *LAC* for the SR 0 and 20% of micro-PnO.

We found that the Dev.% for SR (free Bi_2O_3) is confined between 0.25 and 2.55%, while the Dev.% for the SR with 10 micro-Bi_2O_3 is limited between 0.62 and 3.25%. The Dev.% also ranges between 0.28 and 2.14% for the SR with 20 micro-Bi_2O_3, while the Dev.% for the SR with 30 micro-Bi_2O_3 is limited between 1.22% and 2.54%. These results confirm that the Dev.% is small (less than 3%), which reaffirms the compatibility of the practical and theoretical results.

In Figure 8a, we plot the *LAC* for the SR and the RS with different concentrations of micro-Bi_2O_3 (5, 10, 20 and 30%). Using this figure, we aim to understand the influence of adding some fractions of Bi_2O_3 into the SR on the attenuation performance of the prepared samples. Evidently, the lowest *LAC* is found in the SR and the *LAC* progressively increases as the amount of Bi_2O_3 increases from 5 to 30%, where the maximum *LAC* is reported for the SR + 30% Bi_2O_3 sample. The reason for this enhancement in *LAC* is due to the high density and atomic number of bismuth, and it is known that adding high atomic number elements to the materials increases the probability of the interaction between the photons and the electrons in the materials. Consequently, incorporating Bi_2O_3 into the SR sample leads to the enhancement of the radiation protection performance.

In order to compare the effect of Bi_2O_3 size on the attenuation performance of the SR, we plotted the *LAC* for the SR with 5% micro- and nano-Bi_2O_3 in Figure 8b, and SR with 30% micro- and nano-Bi_2O_3 in Figure 8c. The *LAC* values for the SR-5n is higher than the *LAC* of SR-5m, and the same for 30% (i.e., the *LAC* for the SR with nanoparticles is higher than micro-Bi_2O_3). These results imply that the radiation interaction probability increases when the micro-Bi_2O_3 is replaced with nano-Bi_2O_3 in the SR. From Equation (3), we define a parameter called relative increase (R.I), which shows the enhancement in the *LAC* due to the replacement of micro-Bi_2O_3 by nano-Bi_2O_3. The R.I is higher than 1, which reaffirms the importance of using nanosized Bi_2O_3 to develop an effective attenuation barrier (see Figure 8d). Additionally, the RI for 30% of Bi_2O_3 is higher than that with 5% of Bi_2O_3. Accordingly, SR with 30% of Bi_2O_3 has interesting radiation shielding features, compared to the SR with micro-Bi_2O_3.

Figure 9a,b shows the measured gamma photon transmission through the RS with micro-Bi_2O_3 and nano-Bi_2O_3, respectively. We call this T micro and T nano. It can be seen that both T micro and T nano exponentially decrease with increasing the thickness from 3 to 13 mm. In Figure 9a, the T micro is lower than that of SR (free Bi_2O_3), which means that the transmission of the photons through the sample with Bi_2O_3 is lower than the transmission of photons through the free Bi_2O_3 SR sample. This means that the addition of Bi_2O_3 reduces the transmission of the photons through the prepared silicon rubber.

Additionally, we can see that the T micro depends on the amount of Bi_2O_3 incorporated into the SR. The more Bi_2O_3 in the SR, the lower the T micro. Hence, the incorporation of Bi_2O_3 has a positive influence on the attenuation performance of the SR. If we observe Figure 9b, we can conclude the same results obtained in Figure 9a. In the other words, the photon's transmission through free Bi_2O_3-SR is higher than the SR with nano-Bi_2O_3. This result reaffirms that the increase in the weight fraction of Bi_2O_3 in the SR can efficaciously diminish the photon's transmittance. Therefore, a high amount of Bi_2O_3 (micro or nano sized) in the SR is a good choice to improve the gamma ray shielding performance for the prepared SR.

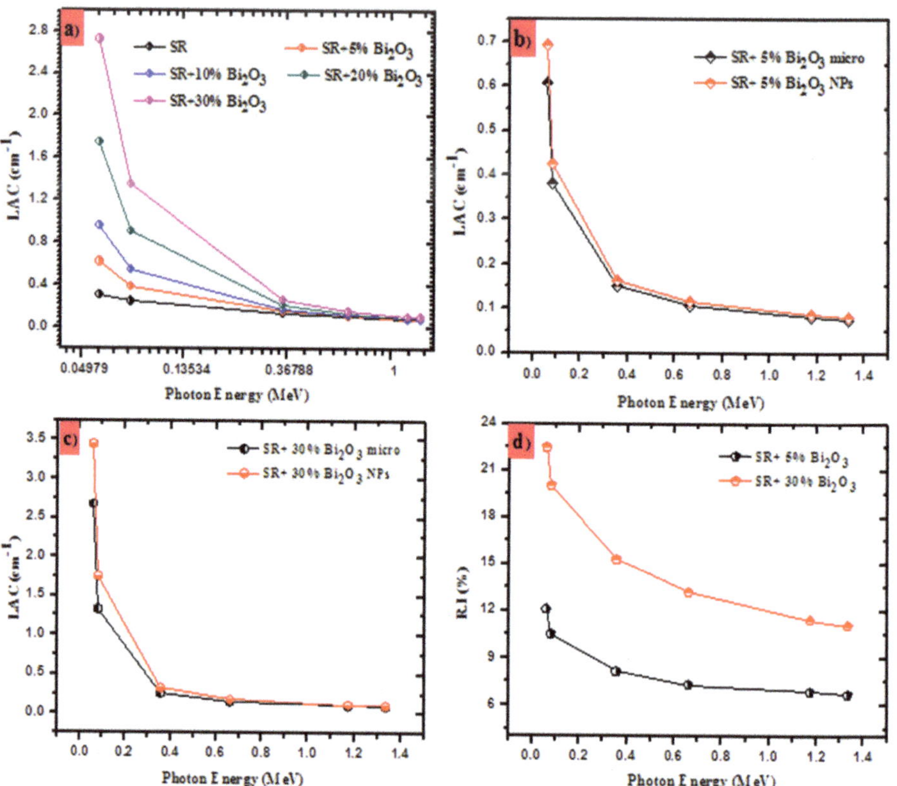

Figure 8. The linear attenuation coefficient for the SR (**a**) with 0, 5, 10, 20 and 30 micro-Bi_2O_3, (**b**) with 5% of micro- and nano-Bi_2O_3, (**c**) with 30% of micro- and nano-Bi_2O_3, and (**d**) the relative increase in the *LAC*.

In order to understand the influence of the thickness of the prepared SR on the attenuation performance, we plotted the radiation shielding efficiency (RPE) for the SR, SR-5m and SR-30m, in Figure 10a, and SR, SR-5n and SR-30n, in Figure 10b, at the same energy value of 0.081 MeV. In both subfigures, it is evident that the RPE for the SR is less than the RPE for the SR with Bi_2O_3 (micro or nano sized), which reaffirms that adding Bi_2O_3 to the SR causes an improvement in the attenuation performance. Most importantly, we can see that the RPE for the SR with a thickness of 13 mm is higher than that with a thickness of 3 mm. This is correct for the SR incorporating micro- or nano-Bi_2O_3. For instance, from Figure 10a, for the SR-5m, the RPE is 10% and this is increased to 40% for the same sample with a thickness of 13 mm. Therefore, we can conclude that the thickness is an important parameter that affects the attenuation competence of the prepared SR. The high thickness SR is recommended as a good attenuator barrier. Moreover, we found that the RPE for the SR with nano-Bi_2O_3 is higher than that of the corresponding micro-Bi_2O_3.

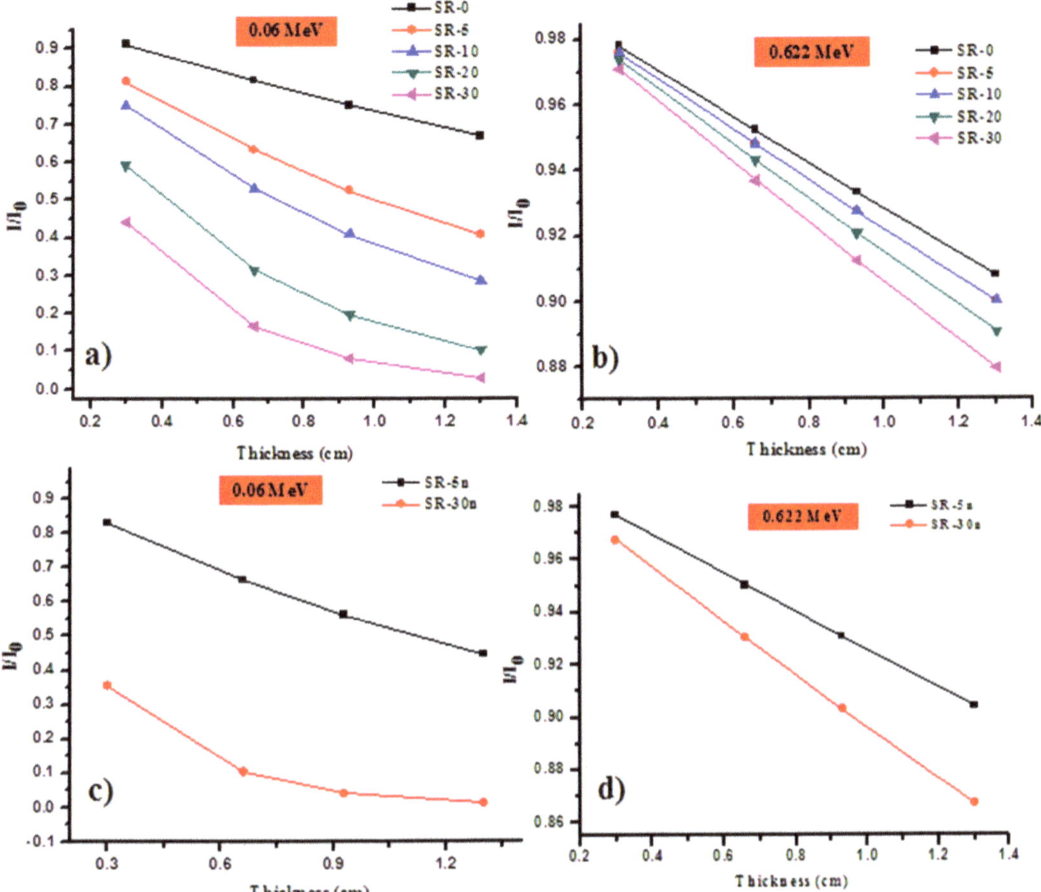

Figure 9. The measured gamma photon transmission through the RS with micro-Bi_2O_3 and nano-Bi_2O_3, respectively, (**a**) RS with micro-Bi_2O_3 at 0.06 MeV, (**b**) RS with micro-Bi_2O_3 at 0.662 MeV, (**c**) RS with nano-Bi_2O_3 at 0.06 MeV and (**d**) RS with nano-Bi_2O_3 at 0.662 MeV.

It is important to compare the radiation attenuation performance for the prepared SR doped with Bi_2O_3 with a similar material, in order to check the possibility of using these samples in real applications. For this purpose, we compared the half value layer of the SR with 30% of micro- and nano-Bi_2O_3 with 3 samples: SR 30, 40 and 50% of magnetite iron [40], as shown in Figure 11. We selected one energy value in the comparison, i.e., 0.662 MeV. Evidently, the SR with 30% of Bi_2O_3 (micro and nano sized) have a lower HVL and thus better attenuation competence than the SR with 30% of magnetite iron. The SR with 30% of micro-Bi_2O_3 has an HVL of 4.52 cm and this is close to 4.62 cm, which was reported for the SR with 40% of iron. The SR with 30% nano-Bi_2O_3 is lower than that of the SR with 30 and 40% of iron, but has an almost similar HVL, with the SR being in contact with 50% of the magnetite iron.

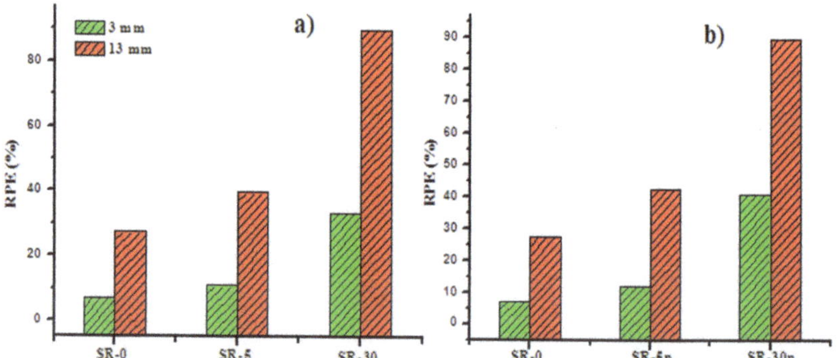

Figure 10. The measured radiation protection efficiency at 3 and 13 mm through (**a**) the RS with micro-Bi_2O_3 and (**b**) nano-Bi_2O_3 at 0.081 MeV.

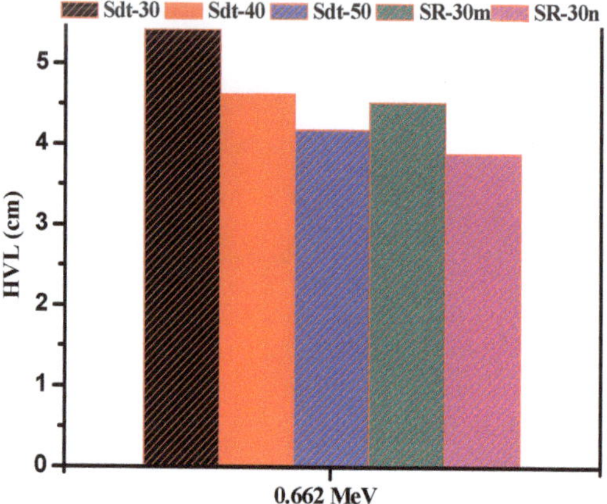

Figure 11. The HVL of 30% micro and nano prepared samples compared with SR filled with magnetite iron.

4. Conclusions

Flexible materials were prepared from SR and different sizes of Bi_2O_3. The morphological, mechanical and shielding properties were determined. The SEM results indicate that the nano filler is significantly better than micro filler. The mechanical results conclude that the flexibility of the materials decreases as we increase the Bi_2O_3 filler with 30 wt%. Therefore, the attenuation study was controlled by the flexibility results. The *LAC* was determined experimentally and the results show good agreement with the theoretical results. The attenuation coefficients of the prepared SR samples showed a clear superiority in lower energy levels over other energies, and the SR's nano-Bi_2O_3 was better than the corresponding SR's micro-Bi_2O_3 at all discussed energies for the shielding materials.

Author Contributions: Conceptualization, M.E. and A.M.E.-K.; methodology, M.E.; software, M.I.S.; validation, H.E.M., M.F.D. and M.I.A.; formal analysis, M.F.D.; investigation, M.I.S.; resources, M.E.; data curation, M.I.A.; writing—original draft preparation, M.F.D.; writing—review and editing, M.I.S.; visualization, A.M.E.-K.; supervision, M.I.A.; project administration, M.E.; funding acquisition, M.F.D. All authors have read and agreed to the published version of the manuscript.

Funding: This research received no external funding.

Institutional Review Board Statement: Not applicable.

Informed Consent Statement: Not applicable.

Data Availability Statement: The data presented in this study are available on request from the corresponding author.

Conflicts of Interest: The authors declare no conflict of interest.

References

1. Dong, M.; Zhou, S.; Xue, X.; Feng, X.; Sayyed, M.I.; Khandaker, M.U.; Bradley, D.A. The potential use of boron containing resources for protection against nuclear radiation. *Radiat. Phys. Chem.* **2021**, *188*, 109601. [CrossRef]
2. Dong, M.; Xue, X.; Yang, H.; Li, Z. Highly cost-effective shielding composite made from vanadium slag and boron-rich slag and its properties. *Radiat. Phys. Chem.* **2017**, *141*, 239–244. [CrossRef]
3. Kamislioglu, M. Research on the effects of bismuth borate glass system on nuclear radiation shielding parameters. *Results Phys.* **2021**, *22*, 103844. [CrossRef]
4. Kamislioglu, M. An investigation into gamma radiation shielding parameters of the (Al:Si) and (Al+Na):Si-doped international simple glasses (ISG) used in nuclear waste management, deploying Phy-X/PSD and SRIM software. *J. Mater. Sci. Mater. Electron.* **2021**, *32*, 12690–12704. [CrossRef]
5. Gökçe, H.; Öztürk, B.C.; Çam, N.; Andiç-Çakır, Ö. Gamma-ray attenuation coefficients and transmission thickness of high consistency heavyweight concrete containing mineral admixture. *Cem. Concr. Compos.* **2018**, *92*, 56–69. [CrossRef]
6. Gökçe, H.S.; Yalçınkaya, Ç.; Tuyan, M. Optimization of reactive powder concrete by means of barite aggregate for both neutrons and gamma rays. *Constr. Build. Mater.* **2018**, *189*, 470–477. [CrossRef]
7. Mahmoud, M.E.; El-Khatib, A.M.; Halbas, A.M.; El-Sharkawy, R.M. Investigation of physical, mechanical and gamma-ray shielding properties using ceramic tiles incorporated with powdered lead oxide. *Ceram. Int.* **2020**, *46*, 15686–15694. [CrossRef]
8. Sayyed, M.I.; Elmahroug, Y.; Elbashir, B.O.; Issa, S. Gamma-ray shielding properties of zinc oxide soda lime silica glasses. *J. Mater. Sci. Mater. Electron.* **2016**, *28*, 4064–4074. [CrossRef]
9. El-Nahal, M.A.; Elsafi, M.; Sayyed, M.I.; Khandaker, M.U.; Osman, H.; Elesawy, B.H.; Saleh, I.H.; Abbas, M.I. Understanding the Effect of Introducing Micro- and Nanoparticle Bismuth Oxide (Bi_2O_3) on the Gamma Ray Shielding Performance of Novel Concrete. *Materials* **2021**, *14*, 6487. [CrossRef]
10. Azeez, A.B.; Kahtan, S.; Mohammed, K.S.; Al Bakrı Abdullah, M.M.; Zulkeplı, N.N.; Sandu, A.V.; Hussın, K.; Rahmat, A. Design of Flexible Green Anti Radiation Shielding Material against Gamma-ray. *Mater. Plast.* **2014**, *51*, 300–308.
11. Yılmaz, S.N.; Gungor, A.; Ozdemir, T. The investigations of mechanical, thermal and rheological properties of polydimethylsiloxane/bismuth (III) oxide composite for X/Gamma ray shielding. *Radiat. Phys. Chem.* **2020**, *170*, 108649. [CrossRef]
12. Mahmoud, M.E.; El-Khatib, A.M.; El-Sharkawy, R.M.; Rashad, A.R.; Badawi, M.S.; Gepreel, M.A. Design and testing of high-density polyethylene nanocompositesfilled with lead oxide micro- and nano-particles: Mechanical, thermal, and morphological properties. *J. Appl. Polym. Sci.* **2019**, *136*, 47812. [CrossRef]
13. Almurayshid, M.; Alsagabi, S.; Alssalim, Y.; Alotaibi, Z.; Almsalam, R. Feasibility of polymer-based composite materials as radiation shield. *Radiat. Phys. Chem.* **2021**, *183*, 109425. [CrossRef]
14. Nagaraja, N.; Manjunatha, H.; Seenappa, L.; Sridhar, K.; Ramalingam, H. Radiation shielding properties of silicon polymers. *Radiat. Phys. Chem.* **2020**, *171*, 108723. [CrossRef]
15. Labouriau, A.; Robison, T.; Shonrock, C.; Simmonds, S.; Cox, B.; Pacheco, A.; Cady, C. Boron filled siloxane polymers for radiation shielding. *Radiat. Phys. Chem.* **2018**, *144*, 288–294. [CrossRef]
16. Ambika, M.; Nagaiah, N.; Harish, V.; Lokanath, N.; Sridhar, M.; Renukappa, N.; Suman, S. Preparation and characterisation of Isophthalic-Bi2O3 polymer composite gamma radiation shields. *Radiat. Phys. Chem.* **2017**, *130*, 351–358. [CrossRef]
17. Karabul, Y.; Içelli, O. The assessment of usage of epoxy based micro and nano-structured composites enriched with Bi_2O_3 and WO_3 particles for radiation shielding. *Results Phys.* **2021**, *26*, 104423. [CrossRef]
18. Kameesy, S.; Nashar, D.; Fiki, S. Development of silicone rubber/lead oxide composites as gamma ray shielding materials. *Int. J. Adv. Res.* **2015**, *3*, 1017–1023. [CrossRef]
19. Gong, P.; Ni, M.; Chai, H.; Chen, F.; Tang, X. Preparation and characteristics of a flexible neutron and γ-ray shielding and radiation-resistant material reinforced by benzophenone. *Nucl. Eng. Technol.* **2018**, *50*, 470–477. [CrossRef]
20. Özdemir, T.; Yılmaz, S.N. Mixed radiation shielding via 3-layered polydimethylsiloxane rubber composite containing hexagonal boron nitride, boron (III) oxide, bismuth (III) oxide for each layer. *Radiat. Phys. Chem.* **2018**, *152*, 17–22. [CrossRef]
21. Chai, H.; Tang, X.; Ni, M.; Chen, F.; Zhang, Y.; Chen, D.; Qiu, Y. Preparation and properties of flexible flame-retardant neutron shielding material based on methyl vinyl silicone rubber. *J. Nucl. Mater.* **2015**, *464*, 210–215. [CrossRef]
22. Colas, A.; Curtis, J. Silicone Biomaterials: History and Chemistry & Medical Applications of Silicones. In *Biomaterials Science*, 2nd ed.; Elsevier Academic Publishing: Amsterdam, The Netherlands, 2004; ISBN 0-12-582463-7.
23. El-Khatib, A.M.; Elsafi, M.; Sayyed, M.; Abbas, M.; El-Khatib, M. Impact of micro and nano aluminium on the efficiency of photon detectors. *Results Phys.* **2021**, *30*, 104908. [CrossRef]
24. Abbas, M.I. Validation of analytical formulae for the efficiency calibration of gamma detectors used in laboratory and in-situ measurements. *Appl. Radiat. Isot.* **2006**, *64*, 1661–1664. [CrossRef] [PubMed]
25. Abbas, M.I. A new analytical method to calibrate cylindrical phoswich and $LaBr_3(Ce)$ scintillation detectors. *Nucl. Instrum. Methods Sect. A* **2010**, *621*, 413–418. [CrossRef]

26. Elsafi, M.; El-Nahal, M.A.; Sayyed, M.I.; Saleh, I.H.; Abbas, M.I. Effect of bulk and nanoparticle Bi_2O_3 on attenuation capability of radiation shielding glass. *Ceram. Int.* **2021**, *47*, 19651–19658. [CrossRef]
27. Eid, M.S.; Bondouk, I.I.; Saleh, H.M.; Omar, K.M.; Sayyed, M.I.; El-Khatib, A.M.; Elsafi, M. Implementation of waste silicate glass into composition of ordinary cement for radiation shielding applications. *Nucl. Eng. Technol.* 2021, *in press*. [CrossRef]
28. Elsafi, M.; Sayyed, M.; Almuqrin, A.H.; Gouda, M.; El-Khatib, A. Analysis of particle size on mass dependent attenuation capability of bulk and nanoparticle PbO radiation shields. *Results Phys.* **2021**, *26*, 104458. [CrossRef]
29. El-Khatib, A.M.; Abbas, M.I.; Elzaher, M.A.; Badawi, M.S.; Alabsy, M.T.; Alharshan, G.A.; Aloraini, D. Gamma Attenuation Coefficients of Nano Cadmium Oxide/High density Polyethylene Composites. *Sci. Rep.* **2019**, *9*, 16012. [CrossRef]
30. Alabsy, M.T.; Alzahrani, J.S.; Sayyed, M.I.; Abbas, M.I.; Tishkevich, D.I.; El-Khatib, A.M.; Elsafi, M. Gamma-Ray Attenuation and Exposure Buildup Factor of Novel Polymers in Shielding Using Geant4 Simulation. *Materials* **2021**, *14*, 5051. [CrossRef]
31. El-Khatib, A.M.; Elsafi, M.; Almutiri, M.N.; Mahmoud, R.M.M.; Alzahrani, J.S.; Sayyed, M.I.; Abbas, M.I. Enhancement of Bentonite Materials with Cement for Gamma-Ray Shielding Capability. *Materials* **2021**, *14*, 4697. [CrossRef]
32. Sayyed, M.I.; Albarzan, B.; Almuqrin, A.H.; El-Khatib, A.M.; Kumar, A.; Tishkevich, D.I.; Trukhanov, A.V.; Elsafi, M. Experimental and Theoretical Study of Radiation Shielding Features of CaO K_2O-Na_2O-P_2O_5 Glass Systems. *Materials* **2021**, *14*, 3772. [CrossRef]
33. Elsafi, M.; Dib, M.F.; Mustafa, H.E.; Sayyed, M.I.; Khandaker, M.U.; Alsubaie, A.; Almalki, A.S.A.; Abbas, M.I.; El-Khatib, A.M. Enhancement of Ceramics Based Red-Clay by Bulk and Nano Metal Oxides for Photon Shielding Features. *Materials* **2021**, *14*, 7878. [CrossRef] [PubMed]
34. Al-Hadeethi, Y.; Sayyed, M.I.; Barasheed, A.Z.; Ahmed, M.; Elsafi, M. Fabrication of Lead Free Borate Glasses Modified by Bismuth Oxide for Gamma Ray Protection Applications. *Materials* **2022**, *15*, 789. [CrossRef]
35. Al-Harbi, N.; Sayyed, M.I.; Al-Hadeethi, Y.; Kumar, A.; Elsafi, M.; Mahmoud, K.A.; Khandaker, M.U.; Bradley, D.A. A novel CaO–K_2O–Na_2O–P_2O_5 glass systems for radiation shielding applications. *Radiat. Phys. Chem.* **2021**, *188*, 109645. [CrossRef]
36. Mhareb, M.H.A.; Zeama, M.; Elsafi, M.; Alajerami, Y.S.; Sayyed, M.I.; Saleh, G.; Hamad, R.M.; Hamad, M.K. Radiation shielding features for various tellurium-based alloys: A comparative study. *J. Mater. Sci. Mater. Electron.* **2021**, *32*, 26798–26811. [CrossRef]
37. Aloraini, D.A.; Almuqrin, A.H.; Sayyed, M.I.; Al-Ghamdi, H.; Kumar, A.; Elsafi, M. Experimental Investigation of Radiation Shielding Competence of Bi_2O_3-CaO-K_2O-Na_2O-P_2O_5 Glass Systems. *Materials* **2021**, *14*, 5061. [CrossRef]
38. Elsafi, M.; Alrashedi, M.; Sayyed, M.; Al-Hamarneh, I.; El-Nahal, M.; El-Khatib, M.; Khandaker, M.; Osman, H.; Askary, A. The Potentials of Egyptian and Indian Granites for Protection of Ionizing Radiation. *Materials* **2021**, *14*, 3928. [CrossRef] [PubMed]
39. Elsafi, M.; Koraim, Y.; Almurayshid, M.; Almasoud, F.I.; Sayyed, M.I.; Saleh, I.H. Investigation of Photon Radiation Attenuation Capability of Different Clay Materials. *Materials* **2021**, *14*, 6702. [CrossRef]
40. Buyuk, B. Gamma-Ray Attenuation Properties of Flexible Silicone Rubber Materials while using Cs-137 as Radioactive Source. *Eur. J. Sci. Technol.* **2019**, *15*, 28–35. [CrossRef]

Article

Compositional Influence on the Morphology and Thermal Properties of Woven Non-Woven Mats of PLA/OLA/MgO Electrospun Fibers

Adrián Leonés [1,2], Laura Peponi [1,2,*], Jesús-María García-Martínez [1,†] and Emilia P. Collar [1,†]

1   Instituto de Ciencia y Tecnología de Polímeros (ICTP-CSIC), C/Juan de la Cierva 3, 28006 Madrid, Spain; aleones@ictp.csic.es (A.L.); jesus.maria@ictp.csic.es (J.-M.G.-M.); ecollar@ictp.csic.es (E.P.C.)

2   Interdisciplinary Platform for "Sustainable Plastics towards a Circular Economy" (SUSPLAST-CSIC), 28006 Madrid, Spain

*   Correspondence: lpeponi@ictp.csic.es

†   Grupo de Ingenieria de Polimeros.

Abstract: In the present work, a statistical study of the morphology and thermal behavior of poly(lactic acid) (PLA)/oligomer(lactic acid) (OLA)/magnesium oxide nanoparticles (MgO), electrospun fibers (efibers) has been carried out. The addition of both, OLA and MgO, is expected to modify the final properties of the electrospun PLA-based nanocomposites for their potential use in biomedical applications. Looking for the compositional optimization of these materials, a Box–Wilson design of experiment was used, taking as dependent variables the average fiber diameter as the representative of the fiber morphologies, as well as the glass transition temperature (T_g) and the degree of crystallinity (X_c) as their thermal response. The results show $<r^2>$ values of 73.76% (diameter), 88.59% (T_g) and 75.61% (X_c) for each polynomial fit, indicating a good correlation between both OLA and MgO, along with the morphological as well as the thermal behavior of the PLA-based efibers in the experimental space scanned.

Keywords: electrospinning; poly(lactic acid); magnesium oxide; oligomer(lactic acid); design of experiments

Citation: Leonés, A.; Peponi, L.; García-Martínez, J.-M.; Collar, E.P. Compositional Influence on the Morphology and Thermal Properties of Woven Non-Woven Mats of PLA/OLA/MgO Electrospun Fibers. *Polymers* **2022**, *14*, 2092. https://doi.org/10.3390/polym14102092

Academic Editor: Lilia Sabantina

Received: 4 April 2022
Accepted: 18 May 2022
Published: 20 May 2022

Publisher's Note: MDPI stays neutral with regard to jurisdictional claims in published maps and institutional affiliations.

1. Introduction

Poly(lactic acid), PLA, is one of the most studied biobased polymers, not only for its biodegradability and renewable source [1], but also for its easy processability [2], low weight [3] and transparency [4]. However, its brittle nature and its low mechanical properties in terms of elongation at break and tensile strength are still drawbacks that need to be improved [5]. Otherwise, PLA shows degradability under physiological conditions into non-toxic products, making it an ideal material to be used in contact with the human body [6,7]. However, considering some biomedical applications, such as films or woven/non-woven electrospun fibers with thermally-activated shape memory behavior for uses in contact with human tissues, its glass transition temperature, T_g, (60 °C) is much higher than the human body temperature, which needs to be tailored [8].

Nowadays, the use of both plasticizers and nanoparticles, NPs, are widely used in order to enhance the mechanical response and modulate the thermal properties of PLA [9,10]. In general, the addition of new components into a polymer strongly affected the final reinforced polymeric matrix behavior [11]; therefore, their interactions have to be studied to guarantee the optimal performance of the material [12]. Moreover, the effect of the different components on the properties of polymer nanocomposites depends on several factors, such as the complete dispersion of the nanoparticles in the matrix and the consequent development of a huge interfacial area [9,11,13,14].

With this background, the ideal plasticizer should have a similar chemical structure to the matrix and appropriate molecular weight to assure the good compatibility between the matrix and the NPs, therefore, enhancing the final thermal and mechanical properties of the nanocomposites [14,15]. In particular for PLA matrix, several additives have been tested as potential plasticizers, such as glycerol [16], poly(ethylene glycol) [17], acetyl(tributyl citrate) [18] or glyceril triacetate [19]. In recent years, an oligomer of lactic acid, OLA, has been proposed as an alternative to common plasticizers for PLA, taking advantage of its similar chemical structure [20–23]. Some studies can be found in the literature in this way. In particular, Cicogna et al. [21] reported the structural, thermal and biodegradability properties of PLA films plasticized with OLA. Additionally, Burgos et al. [20] studied different PLA films plasticized with three OLAs synthesized under different conditions and reported enhanced mechanical properties in terms of elongation at break. The use of OLA as a plasticizer for PLA matrix is not limited to films. In fact, the electrospinning process can be used to obtain PLA plasticized electrospun fibers, efibers. The electrospinning process is the most suitable technique used to produce a wide variety of woven non-woven mats starting from a polymeric solution exposed to high electric fields [24]. Moreover, it is one of the most efficient, simple and versatile processing techniques able to produce fibers for different applications, such as drug delivery [25], food packaging [26], or tissue engineering [27]. The use of OLA as a plasticizer in PLA-based efibers has been recently studied and reported in the literature [8,26,28]. In particular, Leonés et al. [8] obtained PLA-based efibers plasticized with OLA at different concentrations and studied its thermally-activated shape memory behavior tailoring their T_g to a temperature close to the physiological one. In another paper regarding biomedical applications, the in vitro degradation of PLA-based efibers plasticized with OLA has been reported in simulated body fluid until reaching the complete degradation of the samples [28]. They found that the addition of OLA increases the hydrolytic degradation process of PLA-based fiber mats. Moreover, by adding different amounts of OLA, such as 10, 20 and 30 wt%, the time of degradation in phosphate buffered saline, PBS, can be modulated over the course of a year.

On the other hand, the dispersion of different NPs through efibers has been studied with very promising results in different fields, such as food packaging [29], environmental, agricultural [30], and biomedical applications [31]. In particular, in the biomedical field, the main purpose is to produce scaffolds that guide the growth of cells and temporarily serve as a support for cell attachment and differentiation [10]. Thus, scaffolds have to successfully mimic the natural chemical and biological environment and the mechanical properties of the injured tissue [10]. With this aim, different inorganic NPs have been studied to obtain woven non-woven electrospun mats with specific requirements, such as porosity [32], biocompatibility [33], or enhanced mechanical properties [27,34]. Among the inorganic NPs, magnesium oxide nanoparticles, MgO NPs, have been recently studied as reinforcement in polymer matrices due to the role of magnesium in cellular functions, its good biocompatibility and its crucial role in bone growth [10,35]; thus MgO NPs have emerged in recent years as promising inorganic nanofillers for biopolymers [10]. Some authors have recently reported the improved mechanical properties of polymeric efibers reinforced with MgO NPs. For instance, Boakye et al. reported a slight increase in the mechanical properties with the addition of MgO NPs to poly(ε-caprolactone)-keratin efibers [36]. De Silva et al. studied alginate-based efibers reinforced with MgO NPs and reported enhanced tensile strength and elastic modulus [37].

However, no works can be found in the scientific literature about PLA-based efibers with both OLA as a plasticizer and MgO NPs.

Moreover, nowadays, the results of the electrospinning process are still partially unpredictable and exhibit a varying average diameter of efibers which limits its usefulness in industrial applications [38]. At this point, it is important to study the average diameter of the efibers in order to increase the validity and repeatability of the process [39]. In addition, the average diameter is influenced by different parameters involved during the electrospinning process, such as the applied voltage, flow rate, or polymer concentration [10]. It is, therefore, critical to analyze this parameter as representative of the processing of the electrospun fibers.

Therefore, in this work, a preliminary statistical study of the components and their interactions has been carried out in order to study their effects on the PLA matrix. In particular, the main goal of the present study is the optimization of the thermal properties within the experimental range of the composition OLA and MgO for the PLA-based efibers. For this purpose, it is very important to use a response surface methodology, RSM, able to fit any given dependent variables to polynomial equations over the experimental range scanned as a function of those considered as independent ones [40,41]. The RSM encompasses the use of different types of designs of experiments (DOE) which reduce the number of experimental runs necessary to set a reliable mathematical trend within the experimental space scanned for a given response [42,43]. The obtained correlations can be used in further optimization steps. When the limits of the main independent variables are known, the Box–Wilson RSM is one of the most suitable tools for optimization purposes. The Box–Wilson RSM is a central rotatory composite design with ($2^k + 2k + 1$) experiments, plus ($2 + k$) replicated central runs, where k is the number of independent variables [42,43]. Based on previous studies [8,28] a Box–Wilson RSM has been carried out with two independent variables, the amount of OLA and of MgO, considered as the responses to both the morphology and the thermal properties of the efibers. For the former, the evolution of the fiber diameters was chosen, while for the latter they were the glass transition temperature and the degree of crystallinity of the PLA polymer matrix in the nanocomposites. As we will see, very interesting results emerging from the present work include that not only do the two components play an important role in the evolution of the properties of the fiber but there is also a crossed effect linked to the processing step and related to the ratio between both the MgO and the OLA amounts.

2. Materials and Methods

Polylactic acid (PLA3051D), 3% of D-lactic acid monomer, molecular weight 14.2×10^4 g·mol^{-1}, density 1.24 g·cm^{-3}) was supplied by NatureWorks® (NatureWorks LLC, Minnetonka, MN, USA). Lactic acid oligomer (Glyplast OLA8, ester content > 99%, density 1.11 g·cm^{-3}, viscosity 22.5 mPa·s, molecular weight 1100 g·mol^{-1}) was kindly supplied by Condensia Quimica SA (Barcelona, Spain). Chloroform, CHCl$_3$, (99.6% purity) and *N,N*-dimethylformamide, DMF, (99.5% purity) from Sigma Aldrich (Madrid, Spain) were used as solvents. Magnesium oxide nanoparticles (MgO NPs, average particle size of 20 nm, 99.9% purity, molecular weight 40.30 g·mol^{-1}) were supplied by Nanoshel LLC (Wilmington, Delaware, USA).

Previous to the electrospinning process, each solution was prepared following the following steps. Firstly, the corresponding amounts of PLA and OLA were dissolved separately in CHCl$_3$ and stirred overnight at room temperature. Secondly, the amount of MgO NPs was weighed and dispersed in 20 mL of CHCl$_3$; after 30 min the OLA solution was added and dispersed for 60 min. Then, the PLA solution was added and dispersed for another 60 min. Finally, we added the necessary volume of DMF to assure the proportion of solvents CHCl$_3$:DMF (4:1). The dispersion process was carried out with a sonicator tip (Sonic Vibra-Cell VCX 750, Sonics & Materials, Newton, CT, USA) of 750 watts and an amplitude of 20%. Then, electrospun fiber mats were obtained in an Electrospinner Y-flow 2.2.D-XXX (Nanotechnology Solutions, Malaga, Spain) in vertical configuration coupled to coaxial concentric needles. Polymer solutions were pumped through the inner needle and a CHCl$_3$:DMF (4:1) solvent solution was pumped through the outer needle. An electric field of 10 kV in positive and -10 kV in negative poles was set. The flow rates for both the solvent and the polymer solution were fixed at 0.50 mL·h^{-1} and 3.5 mL·h^{-1}, respectively. Each formulation was electrospun for 3 h over a metal plane collector covered with aluminum foil placed at a 14 cm distance from the needle. The obtained mats were vacuum dried for 24 h in order to remove any solvent residues [8].

The different amounts of OLA and MgO studied were set according to the specifications of the Box–Wilson experimental worksheet (Table 1). The range of the independent variable OLA was set from 6.00 to 30.00 wt% while for MgO it was from 0.60 to 3.00 wt% being, respectively, coded as the lowest and the highest (-1, 1). Additionally, the Box–Wilson experimental model considers $\alpha = \sqrt{2}$ as the coded variable for the star points of the worksheet [44,45]. Therefore, in Table 1 all the coded and the controlled factors are listed together with each run number. The corresponding trials were conducted in a randomized way. The variables were statistically analyzed by one-way analysis of variance (ANOVA) and using the statistical computer package Statgraphics Centurion XVII (Statpoint Technologies, Inc., Warrenton,, VA, USA) [44].

Table 1. Worksheet for the Box–Wilson experimental design used.

Run	Coded Factors		Controlled Factors	
	OLA	**MgO**	**OLA (wt%)**	**MgO (wt%)**
I	-1	-1	6.00	0.60
II	1	-1	30.00	0.60
III	-1	1	6.00	3.00
IV	1	1	30.00	3.00
V	$-\sqrt{2}$	0	1.03	1.80
VI	$\sqrt{2}$	0	34.97	1.80
VII	0	$-\sqrt{2}$	18.00	0.10
VIII	0	$\sqrt{2}$	18.00	3.49
IX	0	0	18.00	1.80
X	0	0	18.00	1.80
XI	0	0	18.00	1.80
XII	0	0	18.00	1.80
XIII	0	0	18.00	1.80

Scanning Electron Microscopy, SEM, PHILIPS XL30 Scanning Electron Microscope, (Phillips, Eindhoven, The Netherlands) was used in order to study the morphology of the efibers. All the samples were previously gold-coated ($\sim$5 nm thickness) in a Polaron SC7640 Auto/Manual Sputter (Polaron, Newhaven, East Sussex, UK). SEM image analyses were carried out with ImageJ software (Bethesda, Maryland, USA). Diameters were calculated as the average value of 30 random measurements for each sample.

Thermal transitions were studied by Differential Scanning Calorimetry, DSC, in a DSC Q2000 TA instrument under a nitrogen atmosphere (50 mL·min^{-1}). The samples were cooled from room temperature to -60 °C at 20 °C·min^{-1}, then, the thermal analysis was programmed at 10 °C·min^{-1} from -60 °C up to 180 °C obtaining the glass transition temperature (T_g) calculated as the midpoint of the transition, the cold crystallization enthalpy (ΔH_{cc}) and the melting enthalpy (ΔH_m). The degree of crystallinity ($X_c\%$) was calculated using Equation (1), taking the value of crystallization enthalpy of pure crystalline PLA ($\Delta H_m°$) as 93.6 J·g^{-1} and W_f as the weight fraction of PLA in the sample [45].

$$X_c(\%) = \frac{\Delta H_m - \Delta H_{cc}}{\Delta H_m°} \times \frac{1}{W_f} \times 100 \tag{1}$$

3. Results and Discussion

Once we obtained the different woven non-woven electrospun mats for each run, their fiber morphologies were studied by SEM images. Figure 1 displays typical SEM images of the woven non-woven mats from each one of the runs of Table 1, representing the experimental worksheet. As can be seen, straight and randomly oriented fibers with different average diameters, reported in Table 2, were obtained for each run, which indicates the suitability of the experimental range chosen by using experimental conditions previously reported by us [8,28]. Some beads, small in size, may be observed in the different samples and are considered typical defects when processing by electrospinning, and are associated with solution viscosity and increments in the surface tension [46].

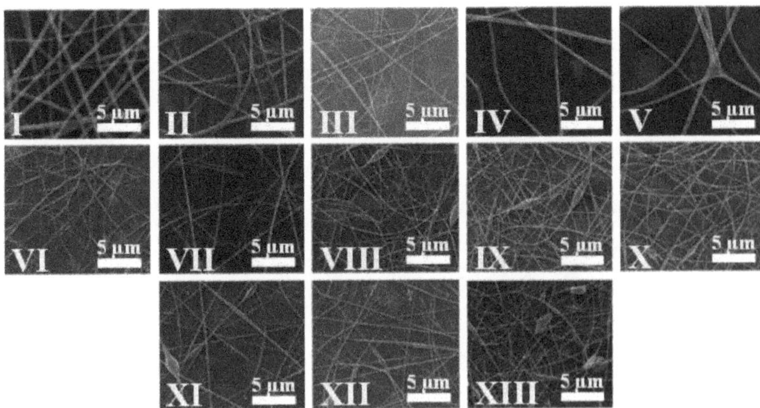

Figure 1. SEM images of the different electrospun mats obtained from run I to run XIII of the experimental worksheet in Table 1.

Table 2. Efibers experimental responses: Average diameter values, glass transition temperature, T_g, and degree of crystallinity, X_c.

Run	Diameter (nm)	T_g (°C)	X_c (%)
I	469 ± 106	57	3.9
II	316 ± 67	31	30.0
III	137 ± 16	56	9.6
IV	156 ± 38	32	17.6
V	438 ± 67	51	3.3
VI	188 ± 45	34	16.6
VII	192 ± 39	43	9.8
VIII	186 ± 56	46	8.4
IX	172 ± 35	42	6.1
X	187 ± 43	44	14.6
XI	205 ± 62	43	12.3
XII	162 ± 42	47	7.7
XIII	163 ± 36	43	8.6

It is possible to note the differences in the diameter between the run V and the run VI SEM images, which are corresponding, respectively, to the lowest and the highest OLA amount in the nanocomposites, and both contain 1.8 wt% of MgO NPs. From a qualitative point of view, the lower diameters in run VI compared to those in run V, agree with the decrease in the viscosity values of the run VI solutions of 35 wt% OLA can be expected, compared to the run V solution with has just 1.03 wt% of OLA.

The quantitative dimension of the morphological study will be given by the polynomial fit to surface response of the average fiber diameter values reported in Table 2, which indicates the glass transition temperature, T_g and the degree of crystallinity, X_c, of PLA, values obtained from DSC thermograms for the different nanocomposites obtained, according to the Box–Wilson worksheet. Moreover, an example of a characteristic DSC thermogram of PLA-based efibers is shown in Figure 2.

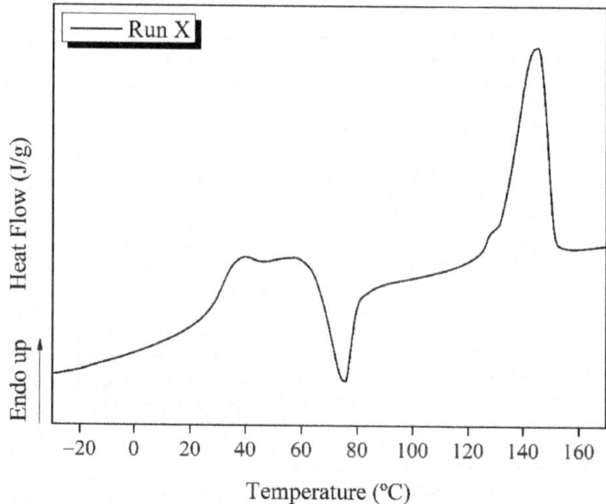

Figure 2. DSC thermogram for PLA-based efibers obtained in run X.

As can be seen in Table 2, all electrospun nanocomposites show a higher degree of crystallinity than the value obtained for the neat PLA efibers, $X_c = 1.2\%$. In particular, runs II, IV and VI show the lowest T_g and are the samples with the highest values for X_c. Run II shows a T_g of 31 °C and X_c of 30%, run IV a T_g of 32 °C and 17.6% and run VI a T_g of 34 °C and X_c of 16.6%, as we expect, lower T_g indicates major mobility of the polymeric chains and as consequence, better facility to crystallize and therefore higher X_c values.

The diameter, T_g and X_c of the PLA nanocomposites were fitted to quadratic models by following the Box–Wilson response surface methodology RSM [42,43]. Therefore, three different polynomials with quadratic and interaction terms were properly obtained having the general form:

$$Y = a_0 + a_1 \cdot x_1 + a_2 \cdot x_2 + a_3 \cdot x_1 \cdot x_2 + a_4 \cdot x_1^2 + a_5 \cdot x_2^2$$

The coefficients obtained for each polynomial fit are reported in Table 3, together with the percentual confidence values for $<r^2>$, the lack of fit and the confidence factors coefficients, obtained from the ANOVA, which informs about the accuracy and significance of the variables.

Table 3. Statistical parameters and coefficients of the polynomial equations from the Box–Wilson experimental design used. ($Y = a_0 + a_1 \cdot x_1 + a_2 \cdot x_2 + a_3 \cdot x_1 \cdot x_2 + a_4 \cdot x_1^2 + a_5 \cdot x_2^2$).

	$<r^2>$ (%)	L. F. * (%)	C. F. * (%)	Ind. T. *	L. T. *		Int. T. *	Q. T. *	
				a_0	a_1	a_2	a_3	a_4	a_5
Diameter (nm)	73.76	1.5	94.1	645.3	−28.51	−131.4	2.986	0.5016	7.101
T_g (°C)	88.59	5.2	99.1	57.86	−0.7364	−1.113	0.03993	−0.003420	0.3004
X_c (%)	75.61	18.9	95.1	−2.816	0.7388	2.096	−0.3142	0.01046	0.7352

* L.F. (Lack of fit), C. F. (Confident Factor), Ind. T. (Independent Term), L. T. (Linear Terms), Int. T. (Interaction Term), Q. T. (Quadratic Terms).

The $<r^2>$ (%) value is the first indicator of how well a model fits a data set and, accordingly, how well a model can predict the value of the response of the variable studied in percentage terms. As reported in the literature [42,43], $<r^2>$ values above 70.00% are considered good fitting for quadratic models. In our case, the $<r^2>$ values obtained were 73.76% for diameter, 88.59% for the T_g and 75.61% for X_c. Therefore, it indicates the significance of OLA and MgO NPs content as the independent variables chosen to model the diameter, T_g and X_c of the PLA-based efibers in our experimental range studied.

Table 3 also shows the lack of fit values associated with the percentage of pure error. In fact, the lack of fit values tells us about possible factors overlooked by the model or a poor choice of variables, but significant in the response development. High values of lack of fit indicate that this parameter is more sensitive to the noise effects of the experiment carried out. As can be seen, lack of fit values of 1.5%, 5.2% and 18.9% were obtained for diameter, T_g and X_c, respectively.

Additionally, high values for the confident factors indicate the full significance of the independent variables chosen in this study. In fact, in our system, confidence factors of 94.1%, 99.1% and 95.1% for diameter, T_g and X_c, respectively, indicate that all the factors considered to build our model play a prime role in the behavior of PLA-based efibers. Consequently, all the parameters summarized in Table 3 confirm the successful choice of studying the system from the Box–Wilson model forecast.

Firstly, the limitations of the model, if any, have to be checked. In this regard, the predicted versus the plot of the measured value is one of the most common alternatives to evaluate models by studying the scatter in a set of data [47].

In Figure 3, in order to explore this aspect, the predicted versus the measured values for diameter, T_g and X_c, are plotted, respectively, in Figure 3a–c. A very good correlation between measured and predicted values and a good scatter of the set of data for each property was observed indicating homoscedastic distribution.

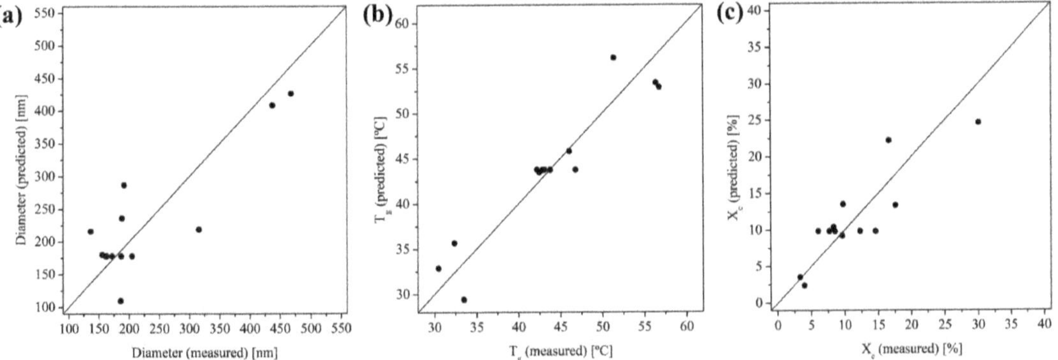

Figure 3. Predicted versus measured values for (**a**) diameter, (**b**) glass transition temperature and (**c**) degree of crystallinity.

Additionally, Table 4 compiles the confidence coefficient (%) and *t*-value for the different terms of each polynomial equation of the Box–Wilson model obtained for the studied properties. In general, the higher the confidence coefficient, the more certain are the results. On the other hand, the *t*-value measures how many standard errors the coefficients are from zero. Generally, any *t*-value higher than 2 is significant and the higher the *t*-value for a term, the greater the confidence in this term [42,43].

We can remark that the diameters show highly significant t-values for both the linear and quadratic terms in the case of OLA which are 3.33 and 2.61, respectively, with very high confidence coefficients, showing a high dependence of this property on OLA content. Additionally, the *t*-value in the limit of significance for the linear parameter in the case of MgO NPs content, 1.54, indicates a slight dependence of diameter on the MgO NPs content. This is also confirmed by its confidence coefficients. A different case relates to the glass transition temperature where only the linear term in the case of OLA, 1.72, can be considered in the limit of significance revealing a dependence of T_g on the plasticizer content. No dependence of T_g on MgO NPs content is significantly observed. In the case of X_c neither the t-values for the linear term, nor for the quadratic terms indicate its dependence on both OLA and MgO content, while the t-value for its interaction terms, 1.96, proves the significance of the interfacial interaction between the plasticizer and the nanoparticles. Scientific implications of these comments will be discussed under the next points, over the corresponding dependent vs independent variable plots.

Table 4. *t*-value and confidence coefficient (%) for the different terms of each polynomial equations of the Box–Wilson model obtained for the studied properties.

	Ind. T. *	L. T. *		Int. T. *	Q. T. *	
		x_1	x_2	$x_1 \cdot x_2$	x_1^2	x_2^2
Diameter (nm)	5.31 (99.7%)	3.33 (98.6%)	1.54 (82.8%)	1.18 (71.0%)	2.61 (96.6%)	0.37 (31.1%)
T_g (°C)	9.51 (99.9%)	1.72 (87.0%)	0.26 (26.0%)	0.31 (28.5%)	0.36 (30.5%)	0.31 (28.4%)
X_c (%)	0.37 (31.0%)	1.37 (77.8%)	0.39 (32.0%)	1.96 (91.1%)	0.86 (56.6%)	0.61 (43.3%)

* Ind. T. (Independent Term), L. T. (Linear Terms), Int. T (Interaction Term), Q. T. (Quadratic Terms).

3.1. Influence of OLA and MgO NPs Content in the Average Diameter of the PLA-Based Efibers

The changes observed in the average diameter evolution in terms of OLA and MgO NPs content have been plotted in Figure 4. In particular, Figure 4a shows the 3D response surface plot for the diameter and Figure 4b shows the contour map properly explaining the evolution of the diameter as a complex function of both OLA and MgO NPs content. First of all, it can be observed that neat PLA efibers show the highest values of diameters with respect to those efibers obtained when both OLA and MgO NPs have been added at different concentrations.

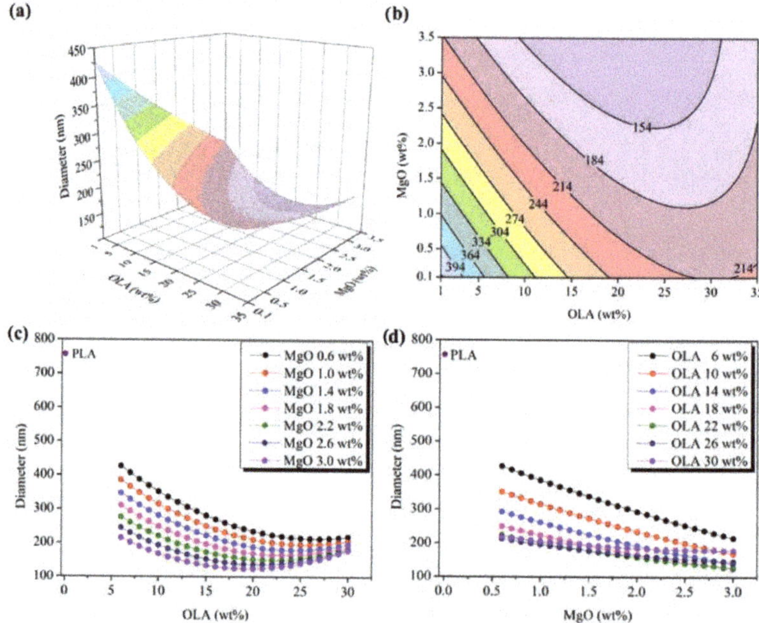

Figure 4. (a) 3D response surface plot and (b) contour plot of the diameter as a function of OLA and MgO NPs content, colours changes are attributed to an increment of 30 nm in the average diameter of efibers. (c) Parametric evolution of diameter with OLA content remaining constant the MgO NPs levels. (d) Parametric evolution of diameter with MgO NPs content remaining constant the OLA amounts.

As described above, the average diameter of the efiber mats is considered representative of studying the processing of the fiber. In general, nanofiller additions tend to increase the viscosity of electrospun polymer solutions [10,48]. However, if nanofillers are conducting materials, such as MgO NPs [49,50], they can strengthen the repulsive force generated during the electrospinning process by increasing the solution conductivity and hence decreasing the fiber diameters. Thus, there is a balance between the viscosity and repulsive force in forming electrospun mats during electrospinning [48].

The parametric plots showing, respectively, the evolution of diameters with OLA at constant levels of MgO NPs, and with the MgO NPs content at constant levels of OLA, have been included in Figure 4c,d. The addition of both, OLA and MgO NPs to PLA, provokes a decrease in the average diameter of the efibers with respect to neat PLA efibers. However, a different evolution of the average diameter can be observed for each independent variable, OLA and MgO NPs. In Figure 4c, the parametric evolution of diameter versus OLA content shows a similar pattern of hyperbola where diameter decreases by increasing the amount of OLA, as expected. However, above 15 wt% of OLA, there is smoothing in the slopes of the hyperbolic curve that reaches up to 25 wt%, where the minimum average diameters can be observed. Finally, from 25 wt% of OLA, the average diameter increases and almost converges at the same value, meaning that at this point, the relation between the increase in viscosity due to the increasing content of MgO NPs is no longer compensated by the increase in the conductivity of the solutions due to the presence of the MgO NPs [47]. Between 17 wt% and 22 wt% of OLA, the effect on the viscosity and on the conductivity due to MgO NPs seems to yield the lowest average diameter of the efibers.

3.2. Influence of OLA and MgO NPs Content in the Glass Transition Temperature of the PLA-Based Efibers

The changes observed in the T_g evolution have been plotted in Figure 5. In particular, Figure 5a shows the 3D response surface plot for T_g and Figure 5b the contour map explaining the evolution of the T_g as a complex function of both OLA and MgO NPs content. Firstly, it is proper to remark that the T_g isolines follow an almost parallel evolution and are hardly influenced by the MgO NPs content up to 2 wt%. In particular, a decrease from 9 to 25 °C by increasing the OLA content in comparison with the T_g of neat PLA (60 °C) in the experimental space scanned, evidenced that the saturation in OLA content is not reached. This behavior is clearly observed in Figure 5c,d, where the parametric evolution of T_g versus OLA and MgO NPs content is shown, respectively.

The parametric evolution of T_g versus OLA content (Figure 5c) evidences the almost linear decreasing dependency of this parameter as the amount of OLA increases, with overlapping for almost all the amount of MgO NPs. In fact, neat PLA efibers show a T_g of 60 °C and the addition of the minimum amount of OLA studied, that is 6 wt%, decreased the T_g value to about 53 °C for both the minimum and maximum amount of MgO NPs studied, that is, 0.6 and 3.0 wt%, respectively. This behavior is observed in all the space scanned until the region of 22 wt% OLA, from this point the isolines begin to separate slightly from each other reaching the maximum amount of OLA studied, that is, 30 wt%, and the distance from each isoline is the highest observed. From the other side, a slight variation in the T_g values can be observed in Figure 5c between the highest and the lowest amount of MgO NPs. In fact, at 30 wt% OLA, when 0.6 MgO wt% has been added, the T_g is 32 °C while when 3 MgO wt% is added, the T_g increases to 35 °C, about 10%. This is a very interesting result that takes into account that usually the addition of NPs at above a minimal concentration leads to an increase in the T_g value of nanocomposites. In particular, Khan et al. reported that at a high volume fraction of NPs, the T_g shows an increasing trend, with an increase in the amount of NPs due to the higher number of contacted polymer chains with NPs and the diffusion of NPs. They concluded that the T_g can be controlled by the agglomeration of NPs and the interaction strength between the NPs and polymer chains [51,52]. In our case, this concentration is not reached in the experimental space scanned and the T_g is only affected by the addition of OLA.

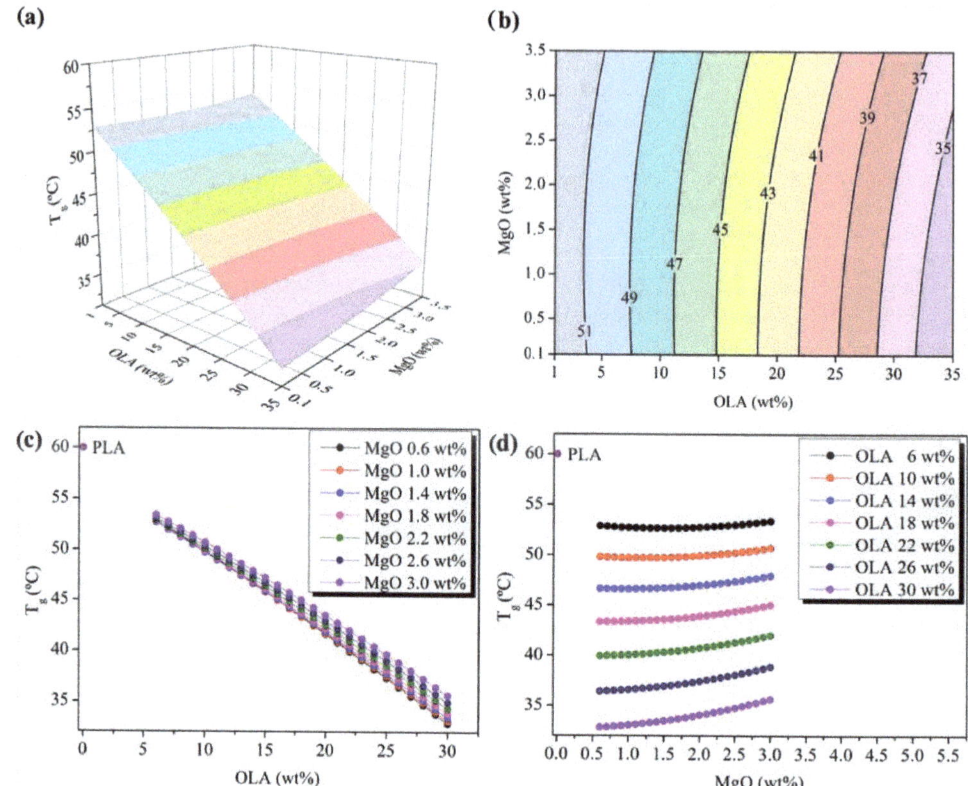

Figure 5. (**a**) 3D response surface plot and (**b**) contour plot of the glass transition temperature as a function of OLA and MgO NPs content, colours changes are attributed to an increment of 2 °C in the T_g. (**c**) Parametric evolution of glass transition temperature with OLA content for different amounts of MgO NPs levels. (**d**) Parametric evolution of glass transition temperature with MgO NPs content for different amounts of OLA amounts.

Therefore, as can be seen in Figure 5c, the T_g of the nanocomposites decreases almost linearly as the OLA content increases, going into the right physiological temperatures window once the 20 wt% OLA content is reached. From here, the overlap observed in the MgO NPs isolines disappears, with those with the highest levels of MgO NPs, 3 wt%, showing the highest T_g. The effect is clearly seen in Figure 5d, where the T_g values evolve almost parallel and equidistant from each other for the lowest OLA and MgO NPs content. A slight increase in the T_g of PLA-reinforced efibers from the values of MgO NPs that are higher than 2.0 wt% is observed.

3.3. Influence of OLA and MgO NPs Content in the Degree of Crystallinity of the PLA-Based Efibers

As described in the literature, the addition of NPs into polymeric matrices is expected to increase the degree of crystallinity due to the nucleation effect of the nanofillers [27,53]. Moreover, some authors have previously reported the capability of OLA to crystallize within the PLA matrix when processed by electrospinning [8,23,28]. Thus, when working with inorganic NPs and OLA plasticizers in PLA-based efibers, the effect of both additives in the X_c has to be considered. In order to study the evolution of this parameter in the experimental space scanned, the response surface plot and the contour map of the degree of crystallinity, calculated by DSC, as a function of OLA and MgO NPs content is reported in Figure 6a,b.

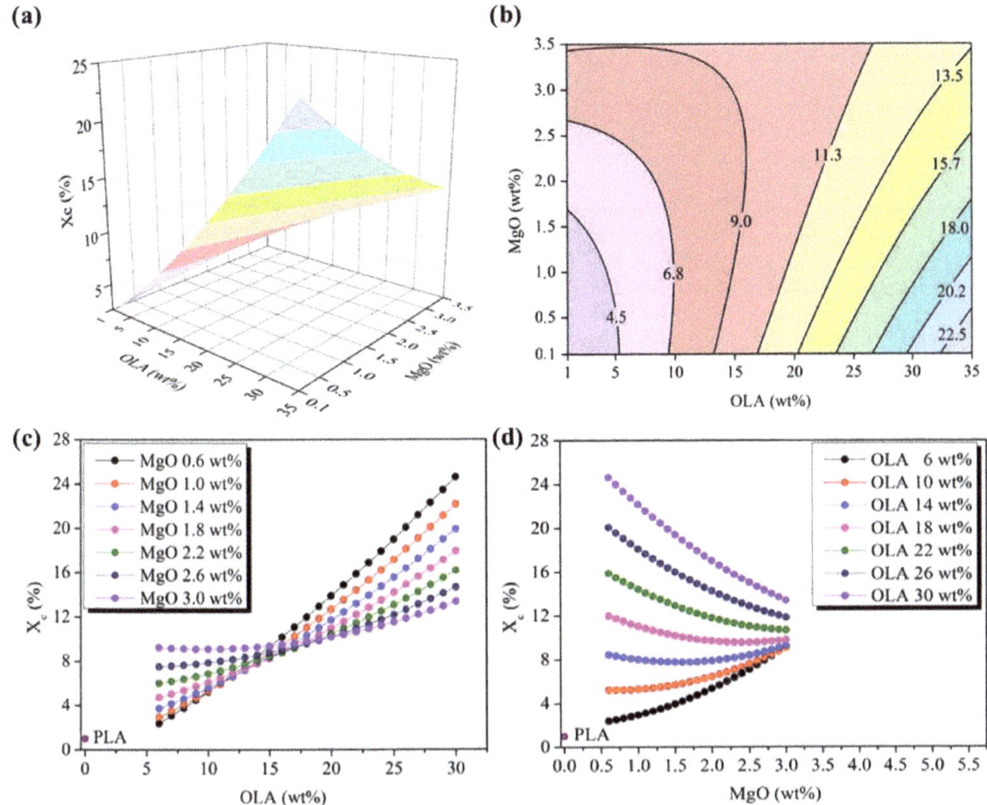

Figure 6. (a) Three-dimensional response surface plot and (b) contour plot of the degree of crystallinity as a function of OLA and MgO NPs content, colours changes are attributed to an increment of 2.3 % in X_c. (c) Parametric evolution of the degree of crystallinity with OLA content for different amounts of MgO NPs levels. (d) Parametric evolution of the degree of crystallinity with MgO NPs content for different amounts of OLA amounts.

As can be observed in Figure 6c, the degree of crystallinity of PLA-based efibers linearly increases with the amount of OLA, showing the highest slopes in the isolines for the lower amount of MgO NPs. This slope progressively smoothens by increasing the amount of MgO NPs until a constant degree of crystallinity of 10% is observed for the highest amount of NPs, that is, MgO NPs 3 wt% (purple isoline in Figure 6c). Moreover, when the amount of OLA reaches 15 wt%, the convergence of the degree of crystallinity takes place for all the MgO NPs content. In particular, from OLA 15 wt%, the degree of crystallinity linearly increases by increasing the amount of OLA with higher slopes for those isolines related to the lower amount of MgO NPs. As can be observed, the highest value of the degree of crystallinity, X_c = 24%, corresponds to the maximum amount of OLA, 30 wt%, and the lowest amount of MgO NPs, which is 0.6 wt%. In fact, it is worth noting, that by increasing the amount of MgO NPs, for the same OLA concentration of 30 wt%, the X_c decreases.

The representation of X_c at the constant levels of OLA shown in Figure 6d shows the same behavior. The highest degrees of crystallinity are observed for the highest amounts of OLA. Moreover, the X_c isolines decrease by increasing the MgO NPs content, smoothing progress on the slopes until reaching the isoline of 18 wt% OLA (pink isoline in Figure 6d). Above 2 wt% MgO NPs, the inversion in the direction of the slopes occurs, which becomes positive for the lower amounts of OLA, showing a degree of crystallinity values towards $X_c = 10\%$.

Some authors previously studied the synergic effects between other nucleating agents and plasticizers, even at very small concentrations, on the degree of crystallinity of PLA nanocomposites, suggesting that, given the enhanced mobility of plasticized PLA, very small concentrations of NPs are effective nucleation agents. They reported a critical amount where a strong effect on the crystallization process was observed and which increases the mechanical properties [54,55]. In our PLA-based efibers, the detection of a region of convergence in a certain region of OLA and MgO NPs content, is coincident with those previously detected for the evolution of the glass transition temperature and diameter, again suggesting a critical MgO/OLA ratio related to the processing stage and determinant in the behavior characteristics of the nanocomposites. The mechanical behavior of these PLA-based efibers will be discussed in upcoming studies, as well as in subsequent studies regarding the design of the experimentation.

4. Conclusions

The role of the addition of both OLA as a plasticizer and MgO as a nanofiller into electrospun PLA fibers has been determined by studying the average diameter of fibers, their glass transition temperature and the degree of crystallinity by statistical analysis. In our electrospun nanocomposites, the interactions between the components determine the final properties and a Box–Wilson model has been used in order to identify the level of interactions and to look for the optimal compositional ratios. The $<r^2>$ values obtained were 73.76% (diameter), 88.59% (T_g) and 75.61% (X_c), respectively, and considered good fitting for quadratic models. In consequence, the predicted versus the measured values for diameter, T_g and X_c, showed a very good correlation evolving as homoscedastic distributions.

The interval of OLA content which leads to a minimum average diameter of fibers has been identified; the results obtained for the glass transition temperature evolution evidence a clear dependency of this parameter on the amount of OLA, whose parametric isolines are almost overlapping for all the amount of MgO NPs. Finally, the synergic effect of OLA and MgO NPs in the degree of crystallinity of the PLA-based efibers increases this parameter from almost zero in the pristine PLA efibers to a range values from 2–24%. A critical point in the X_c evolution has been identified whose coordinates are fully coincident with the critical ones for the other studied properties.

Author Contributions: Conceptualization, L.P., J.-M.G.-M. and E.P.C.; methodology, J.-M.G.-M. and E.P.C.; software, J.-M.G.-M. and A.L.; validation, L.P.; investigation, A.L.; resources, L.P. data curation, J.-M.G.-M.; writing—original draft preparation, A.L.; writing—review and editing, A.L., L.P., J.-M.G.-M. and E.P.C.; supervision, L.P.; project administration, L.P.; funding acquisition, L.P. All authors have read and agreed to the published version of the manuscript.

Funding: Part of this work was financed with the support of the Spanish Ministry of Science and Innovation through MAT2017-88123-P.

Institutional Review Board Statement: Not applicable.

Informed Consent Statement: Not applicable.

Data Availability Statement: Not applicable.

Acknowledgments: The authors want to thank Condensia Quimica S.A. for providing Glyplast OLA8.

Conflicts of Interest: The authors declare no conflict of interest.

References

1. Singhvi, M.; Gokhale, D. Biomass to biodegradable polymer (PLA). *RSC Adv.* **2013**, *3*, 13558. [CrossRef]
2. Nofar, M.; Sacligil, D.; Carreau, P.J.; Kamal, M.R.; Heuzey, M.-C. Poly (lactic acid) blends: Processing, properties and applications. *Int. J. Biol. Macromol.* **2019**, *125*, 307–360. [CrossRef] [PubMed]
3. Rhim, J.-W.; Park, H.-M.; Ha, C.-S. Bio-nanocomposites for food packaging applications. *Prog. Polym. Sci.* **2013**, *38*, 1629–1652. [CrossRef]

4. Raquez, J.M.; Habibi, Y.; Murariu, M.; Dubois, P. Polylactide (PLA)-based nanocomposites. *Prog. Polym. Sci.* **2013**, *38*, 1504–1542. [CrossRef]

5. Farah, S.; Anderson, D.G.; Langer, R. Physical and mechanical properties of PLA, and their functions in widespread applications—A comprehensive review. *Adv. Drug Deliv. Rev.* **2016**, *107*, 367–392. [CrossRef]

6. da Silva, D.; Kaduri, M.; Poley, M.; Adir, O.; Krinsky, N.; Shainsky-Roitman, J.; Schroeder, A. Biocompatibility, biodegradation and excretion of polylactic acid (PLA) in medical implants and theranostic systems. *Chem. Eng. J.* **2018**, *340*, 9–14. [CrossRef]

7. Zaaba, N.F.; Jaafar, M. A review on degradation mechanisms of polylactic acid: Hydrolytic, photodegradative, microbial, and enzymatic degradation. *Polym. Eng. Sci.* **2020**, *60*, 2061–2075. [CrossRef]

8. Leonés, A.; Sonseca, A.; López, D.; Fiori, S.; Peponi, L. Shape memory effect on electrospun PLA-based fibers tailoring their thermal response. *Eur. Polym. J.* **2019**, *117*, 217–226. [CrossRef]

9. Darie-Niţă, R.N.; Vasile, C.; Irimia, A.; Lipşa, R.; Râpă, M. Evaluation of some eco-friendly plasticizers for PLA films processing. *J. Appl. Polym. Sci.* **2016**, *133*, 43223. [CrossRef]

10. Leonés, A.; Lieblich, M.; Benavente, R.; Gonzalez, J.L.; Peponi, L. Potential applications of magnesium-based polymeric nanocomposites obtained by electrospinning technique. *Nanomaterials* **2020**, *10*, 1524. [CrossRef]

11. Peponi, L.; Puglia, D.; Torre, L.; Valentini, L.; Kenny, J.M. Processing of nanostructured polymers and advanced polymeric based nanocomposites. *Mater. Sci. Eng. R. Rep.* **2014**, *85*, 1–46. [CrossRef]

12. García-Martínez, J.M.; Areso, S.; Collar, E.P. The role of a novel p-phenylen-bis-maleamic acid grafted atactic polypropylene interfacial modifier in polypropylene/mica composites as evidenced by tensile properties. *J. Appl. Polym. Sci.* **2009**, *113*, 3929–3943. [CrossRef]

13. Nofar, M.; Salehiyan, R.; Ray, S.S. Influence of nanoparticles and their selective localization on the structure and properties of polylactide-based blend nanocomposites. *Compos. Part B Eng.* **2021**, *215*, 108845. [CrossRef]

14. Rahman, M.; Brazel, C.S. The plasticizer market: An assessment of traditional plasticizers and research trends to meet new challenges. *Prog. Polym. Sci.* **2004**, *29*, 1223–1248. [CrossRef]

15. Bocqué, M.; Voirin, C.; Lapinte, V.; Caillol, S.; Robin, J.-J. Petro-based and bio-based plasticizers: Chemical structures to plasticizing properties. *J. Polym. Sci. Part A Polym. Chem.* **2016**, *54*, 11–33. [CrossRef]

16. Satriyatama, A.; Rochman, V.A.A.; Adhi, R.E. Study of the Effect of Glycerol Plasticizer on the Properties of PLA/Wheat Bran Polymer Blends. *IOP Conf. Ser. Mater. Sci. Eng.* **2021**, *1143*, 012020. [CrossRef]

17. Pivsa-Art, W.; Fujii, K.; Nomura, K.; Aso, Y.; Ohara, H.; Yamane, H. The effect of poly(ethylene glycol) as plasticizer in blends of poly(lactic acid) and poly(butylene succinate). *J. Appl. Polym. Sci.* **2016**, *133*, 43044. [CrossRef]

18. Arrieta, M.P.; Peponi, L.; López, D.; Fernández-García, M. Recovery of yerba mate (Ilex paraguariensis) residue for the development of PLA-based bionanocomposite films. *Ind. Crops Prod.* **2018**, *111*, 317–328. [CrossRef]

19. Salas-Papayanopolos, H.; Morales-Cepeda, A.B.; Sanchez, S.; Lafleur, P.G.; Gomez, I. Synergistic effect of silver nanoparticle content on the optical and thermo-mechanical properties of poly(l-lactic acid)/glycerol triacetate blends. *Polym. Bull.* **2017**, *74*, 4799–4814. [CrossRef]

20. Burgos, N.; Tolaguera, D.; Fiori, S.; Jiménez, A. Synthesis and Characterization of Lactic Acid Oligomers: Evaluation of Performance as Poly(Lactic Acid) Plasticizers. *J. Polym. Environ.* **2014**, *22*, 227–235. [CrossRef]

21. Cicogna, F.; Coiai, S.; De Monte, C.; Spiniello, R.; Fiori, S.; Franceschi, M.; Braca, F.; Cinelli, P.; Fehri, S.M.K.; Lazzeri, A.; et al. Poly(lactic acid) plasticized with low-molecular-weight polyesters: Structural, thermal and biodegradability features. *Polym. Int.* **2017**, *66*, 761–769. [CrossRef]

22. Burgos, N.; Martino, V.P.; Jiménez, A. Characterization and ageing study of poly(lactic acid) films plasticized with oligomeric lactic acid. *Polym. Degrad. Stab.* **2013**, *98*, 651–658. [CrossRef]

23. Arrieta, M.P.; López, J.; López, D.; Kenny, J.M.; Peponi, L. Development of flexible materials based on plasticized electrospun PLA-PHB blends: Structural, thermal, mechanical and disintegration properties. *Eur. Polym. J.* **2015**, *73*, 433–446. [CrossRef]

24. Khorshidi, S.; Solouk, A.; Mirzadeh, H.; Mazinani, S.; Lagaron, J.M.; Sharifi, S.; Ramakrishna, S. A review of key challenges of electrospun scaffolds for tissue-engineering applications. *J. Tissue Eng. Regen. Med.* **2016**, *10*, 715–738. [CrossRef]

25. Basar, A.O.; Castro, S.; Torres-Giner, S.; Lagaron, J.M.; Turkoglu Sasmazel, H. Novel poly(ε-caprolactone)/gelatin wound dressings prepared by emulsion electrospinning with controlled release capacity of Ketoprofen anti-inflammatory drug. *Mater. Sci. Eng. C* **2017**, *81*, 459–468. [CrossRef]

26. Marina P Arrieta; Miguel Perdiguero; Stefano Fiori; José M Kenny; Laura Peponi Biodegradable electrospun PLA-PHB fibers plasticized with oligomeric lactic acid. *Polym. Degrad. Stab.* **2020**, *179*, 109226.

27. Leonés, A.; Mujica-Garcia, A.; Arrieta, M.P.; Salaris, V.; Lopez, D.; Kenny, J.M.; Peponi, L. Organic and Inorganic PCL-Based Electrospun Fibers. *Polymers* **2020**, *12*, 1325. [CrossRef]

28. Leonés, A.; Peponi, L.; Lieblich, M.; Benavente, R.; Fiori, S. In vitro degradation of plasticized PLA electrospun fiber mats: Morphological, thermal and crystalline evolution. *Polymers* **2020**, *12*, 2975. [CrossRef]

29. Rodríguez-Sánchez, I.J.; Fuenmayor, C.A.; Clavijo-Grimaldo, D.; Zuluaga-Domínguez, C.M. Electrospinning of ultra-thin membranes with incorporation of antimicrobial agents for applications in active packaging: A review. *Int. J. Polym. Mater. Polym. Biomater.* **2021**, *70*, 1053–1076. [CrossRef]

30. Arrieta, M.P.; López, J.; López, D.; Kenny, J.M.; Peponi, L. Effect of chitosan and catechin addition on the structural, thermal, mechanical and disintegration properties of plasticized electrospun PLA-PHB biocomposites. *Polym. Degrad. Stab.* **2016**, *132*, 145–156. [CrossRef]

31. Dziemidowicz, K.; Sang, Q.; Wu, J.; Zhang, Z.; Zhou, F.; Lagaron, J.M.; Mo, X.; Parker, G.J.M.; Yu, D.-G.; Zhu, L.-M.; et al. Electrospinning for healthcare: Recent advancements. *J. Mater. Chem. B* **2021**, *9*, 939–951. [CrossRef] [PubMed]

32. Mikos, A.G.; Temenoff, J.S. Formation of highly porous biodegradable scaffolds for tissue engineering. *Electron. J. Biotechnol.* **2000**, *3*, 1995–2000. [CrossRef]

33. Zhu, G.; Zhang, T.; Chen, M.; Yao, K.; Huang, X.; Zhang, B.; Li, Y.; Liu, J.; Wang, Y.; Zhao, Z. Bone physiological microenvironment and healing mechanism: Basis for future bone-tissue engineering scaffolds. *Bioact. Mater.* **2021**, *6*, 4110–4140. [CrossRef] [PubMed]

34. Leonés, A.; Salaris, V.; Mujica-Garcia, A.; Arrieta, M.P.; Lopez, D.; Lieblich, M.; Kenny, J.M.; Peponi, L. PLA Electrospun Fibers Reinforced with Organic and Inorganic Nanoparticles: A Comparative Study. *Molecules* **2021**, *26*, 4925. [CrossRef] [PubMed]

35. Ferrández-Montero, A.; Lieblich, M.; González-Carrasco, J.L.; Benavente, R.; Lorenzo, V.; Detsch, R.; Boccaccini, A.R.; Ferrari, B. Development of biocompatible and fully bioabsorbable PLA/Mg films for tissue regeneration applications. *Acta Biomater.* **2019**, *98*, 114–124. [CrossRef] [PubMed]

36. Boakye, M.; Rijal, N.; Adhikari, U.; Bhattarai, N. Fabrication and Characterization of Electrospun PCL-MgO-Keratin-Based Composite Nanofibers for Biomedical Applications. *Materials* **2015**, *8*, 4080–4095. [CrossRef]

37. De Silva, R.T.; Mantilaka, M.M.M.G.P.G.; Goh, K.L.; Ratnayake, S.P.; Amaratunga, G.A.J.; de Silva, K.M.N. Magnesium Oxide Nanoparticles Reinforced Electrospun Alginate-Based Nanofibrous Scaffolds with Improved Physical Properties. *Int. J. Biomater.* **2017**, *2017*, 1–9. [CrossRef]

38. Cramariuc, B.; Cramariuc, R.; Scarlet, R.; Manea, L.R.; Lupu, I.G.; Cramariuc, O. Fiber diameter in electrospinning process. *J. Electrostat.* **2013**, *71*, 189–198. [CrossRef]

39. El-hadi, A.; Al-Jabri, F. Influence of Electrospinning Parameters on Fiber Diameter and Mechanical Properties of Poly(3-Hydroxybutyrate) (PHB) and Polyanilines (PANI) Blends. *Polymers* **2016**, *8*, 97. [CrossRef]

40. Martínez, J.M.G.; Collar, E.P. On the combined effect of both the reinforcement and a waste based interfacial modifier on the matrix glass transition in ipp/a-pp-ppbma/mica composites. *Polymers* **2020**, *12*, 1–13.

41. García-Martínez, J.-M.; Collar, E.P. The Role of a Succinyl Fluorescein-Succinic Anhydride Grafted Atactic Polypropylene on the Dynamic Mechanical Properties of Polypropylene/Polyamide-6 Blends at the Polypropylene Glass Transition. *Polymers* **2020**, *12*, 1216. [CrossRef]

42. Fisher, R.A. *The Desing of Experiments*, 1st ed.; Hafner: New York, NY, USA, 1960.

43. Box, G.E.P.; Hunter, W.G.; Hunter, J.S. Response surface methods. In *Statistics for Experimenters*, 1st ed.; Wiley & Sons: New York, NY, USA, 1978.

44. Arrieta, M.P.; Gil, A.L.; Yusef, M.; Kenny, J.M.; Peponi, L. Electrospinning of PCL-based blends: Processing optimization for their scalable production. *Materials* **2020**, *13*, 3853. [CrossRef]

45. Peponi, L.; Navarro-Baena, I.; Báez, J.E.; Kenny, J.M.; Marcos-Fernández, A. Effect of the molecular weight on the crystallinity of PCL-b-PLLA di-block copolymers. *Polymer* **2012**, *53*, 4561–4568. [CrossRef]

46. Manea, L.R.; Bertea, A.-P.; Nechita, E.; Popescu, C.V. Mathematical Modeling of the Relation between Electrospun Nanofibers Characteristics and the Process Parameters. In *Electrospinning Method Used to Create Functional Nanocomposites Films*; InTech: London, UK, 2018; p. 13.

47. Piñeiro, G.; Perelman, S.; Guerschman, J.P.; Paruelo, J.M. How to evaluate models: Observed vs. predicted or predicted vs. observed? *Ecol. Modell.* **2008**, *216*, 316–322. [CrossRef]

48. Liu, C.; Shen, J.; Liao, C.Z.; Yeung, K.W.K.; Tjong, S.C. Novel electrospun polyvinylidene fluoride-graphene oxide-silver nanocomposite membranes with protein and bacterial antifouling characteristics. *Express Polym. Lett.* **2018**, *12*, 365–382. [CrossRef]

49. Kathrein, H.; Freund, F. Electrical conductivity of magnesium oxide single crystal below 1200 K. *J. Phys. Chem. Solids* **1983**, *44*, 177–186. [CrossRef]

50. Habeeb, M.A.; Hamza, R.S.A. Synthesis of (Polymer blend-MgO) nanocomposites and studying electrical properties for piezo-electric application. *Indones. J. Electr. Eng. Inform.* **2018**, *6*, 428–435.

51. Khan, R.A.A.; Qi, H.-K.; Huang, J.-H.; Luo, M.-B. A simulation study on the effect of nanoparticle size on the glass transition temperature of polymer nanocomposites. *Soft Matter* **2021**, *17*, 8095–8104. [CrossRef]

52. Serenko, O.A.; Roldughin, V.I.; Askadskii, A.D.; Serkova, E.S.; Strashnov, P.V.; Shifrina, Z.B. The effect of size and concentration of nanoparticles on the glass transition temperature of polymer nanocomposites. *RSC Adv.* **2017**, *7*, 50113–50120. [CrossRef]

53. Papageorgiou, G.Z.; Achilias, D.S.; Bikiaris, D.N.; Karayannidis, G.P. Crystallization kinetics and nucleation activity of filler in polypropylene/surface-treated SiO2 nanocomposites. *Thermochim. Acta* **2005**, *427*, 117–128. [CrossRef]

54. Shi, X.; Zhang, G.; Phuong, T.V.; Lazzeri, A. Synergistic effects of nucleating agents and plasticizers on the crystallization behavior of Poly(lactic acid). *Molecules* **2015**, *20*, 1579–1593. [CrossRef]

55. Clarkson, C.M.; El Awad Azrak, S.M.; Schueneman, G.T.; Snyder, J.F.; Youngblood, J.P. Crystallization kinetics and morphology of small concentrations of cellulose nanofibrils (CNFs) and cellulose nanocrystals (CNCs) melt-compounded into poly(lactic acid) (PLA) with plasticizer. *Polymer* **2020**, *187*, 122101. [CrossRef]

 *polymers*

Article

Combined Effects from Dual Incorporation of ATBC as Plasticizer and Mesoporous MCM-41 as Nucleating Agent on the PLA Isothermal Crystallization in Environmentally-Friendly Ternary Composite Systems

Enrique Blázquez-Blázquez, Rosa Barranco-García, Tamara M. Díez-Rodríguez, María L. Cerrada * and Ernesto Pérez

Instituto de Ciencia y Tecnología de Polímeros (ICTP-CSIC), Juan de la Cierva 3, 28006 Madrid, Spain
* Correspondence: mlcerrada@ictp.csic.es; Tel.: +34-91-258-7474

Abstract: Different materials, based on an *L*-rich polylactide (PLA) as matrix, acetyl tri-*n*-butyl citrate (ATBC) as plasticizer, and mesoporous Mobile Crystalline Material.41 (MCM-41) particles as nucleating agent, were attained by melt extrusion. These materials are constituted by (a) binary blends of PLA and ATBC with different contents of the latest; (b) a dual compound of PLA and a given amount of MCM-41 silica (5 wt.%); and (c) ternary composites that include PLA, ATBC at several compositions and mesoporous MCM-41 at 5 wt.%. Influence of the incorporation of the plasticizer and nucleating particles has been comprehensively analyzed on the different phase transitions: glass transition, cold crystallization, melt crystallization and melting processes. Presence of both additives moves down the temperature at which PLA phase transitions take place, while allowing the PLA crystallization from the melt at 10 °C/min in the composites. This tridimensional ordering is not noticeable in the pristine PLA matrix and, accordingly, PLA crystallization rate is considerably increased under dynamic conditions and also after isothermal crystallization from either the melt or the glassy state. An important synergistic effect of dual action of ATBC and MCM-41 has been, therefore, found.

Keywords: PLA; ATBC; MCM-41; plasticizer; nucleating agent; crystallization rate; glass transition; crystalline polymorphs

Citation: Blázquez-Blázquez, E.; Barranco-García, R.; Díez-Rodríguez, T.M.; Cerrada, M.L.; Pérez, E. Combined Effects from Dual Incorporation of ATBC as Plasticizer and Mesoporous MCM-41 as Nucleating Agent on the PLA Isothermal Crystallization in Environmentally-Friendly Ternary Composite Systems. *Polymers* **2023**, *15*, 624. https://doi.org/10.3390/polym15030624

Academic Editor: Jose-Ramon Sarasua

Received: 14 December 2022
Revised: 16 January 2023
Accepted: 19 January 2023
Published: 25 January 2023

1. Introduction

Poly(lactide) (PLA), a biodegradable aliphatic polyester with an excellent profile in properties, could play a major role in future polymers markets for numerous applications because its monomer comes from renewable resources. PLA shows, however, some shortcomings, such as its brittleness or its relatively reduced crystallization ability under common processing conditions [1]. The former is ascribed to location of its glass transition temperature (T_g), which appears at around 60 °C, and the latest is associated with its low crystallization rate. The position of T_g can be easily adjusted by adding plasticizers, while its three-dimensional ordering capacity can be modulated either by incorporating nucleating agents, since they contribute to lowering the activation energy of nucleation, or by adding plasticizers, due to the increase in the overall mobility of polymeric chains that their presence plays. Nevertheless, the joint incorporation of these additives, plasticizer and nucleanting agent, is an approach little explored in literature [2], although it can be quite attractive from academic and practical standpoints. On one hand, presence of the former can reduce the significant and negative impact that the physical aging, commonly observed in PLA, exerts in its behavior, which promotes a deficient dimensional stability and changes in other key parameters. On the other hand, addition of a nucleant agent will boost crystallization capability of PLA and its rate, inducing the enhancement of some important properties, such as overall strength, thermal resistance or gas barrier. Accordingly, a favorable dual action of both would lead to a balanced and improved response in either amorphous or crystalline phases within the PLA matrix and, subsequently, in the ultimate performance of the resultant ternary materials.

Plasticizers are widely used to enhance flexibility, processability and ductility of polymers. Nevertheless, different variables must be taken into consideration for achieving an efficient effect, which include the nature of matrix, the type and optimal percentage of plasticizer, its thermal stability at the processing temperature, among others. In semicrystalline polymers, such as PLA, incorporation of a plasticizer should reduce the T_g and, additionally, allow the shift to lower temperatures of the crystallization window. The plasticizer efficiency is also noticeable through a melting point depression of the corresponding crystalline phase [3,4]. Numerous types of plasticizers of low molecular weight have been studied for PLA, such as the bishydroxymethyl malonate (DBM) [5], the glucose monoesters and partial fatty acid esters, the citrates [6], among others. Nonetheless, they show often the problem of migration, due to their high mobility within the PLA macrochains. Thus, plasticizers with higher molecular weight and, accordingly, lower mobility have been also evaluated, such as the poly(ethylene glycol) (PEG) [6–8], the poly(propylene glycol) (PPG) [9], the atactic poly(3-hydroxybutyrate) (a-PHB) [10,11] and the oligoesteramide (DBM-oligoesteramide) [5,12,13]. Furthermore, the plasticizer selection for PLA is also restricted by the legislative or technical requirements of the final applications if these are, for instance, as food packaging or in the biomedical field. Therefore, the choice becomes even more complex [5,13]. The good solubility of PLA in the citrate plasticizers, owing to the polar interactions between the ester groups of PLA and the plasticizer [14], makes the acetyl tri-*n*-butyl citrate (ATBC), which is also biodegradable and non-toxic compound, a successful and common alternative to be added to PLA [1,4,13].

Global rates of the nucleation and the crystallization in the PLA under homogeneous conditions are, as aforementioned, relatively low. Accordingly, the addition of nucleating agents is a strategy to promote both processes. These additives allow increasing the number of primary nucleation sites and reducing the nucleation induction period. Different physical agents have been analyzed in PLA for this role, including organic, mineral and mineral-organic hybrids [15]. Talc has been commonly considered as a reference in PLA to evaluate the nucleation capacity of other agents [15,16]. Carbon nanotubes, either neat or modified, have been also employed [15] due to their high aspect ratio and excellent mechanical, thermal and electrical properties. Recently, the mesoporous silica has been used as well. In particular, the SBA-15 particles have been proved to play an important role in the PLA crystallization in composites prepared either by melt extrusion [17,18] or by solution mixing [19]. Mesoporous SBA-15 is constituted by ordered hexagonal arrangements within the particles, which are developed during the synthesis and consist in hollow channels with diameters from around 7 to 10 nm [20].

The aim of this research is to evaluate the joint influence of the dual incorporation of a plasticizer and a mesoporous silica in the crystallization process of PLA. Then, ATBC has been chosen for the former and MCM-41 as silica in the preparation of ternary systems. This plasticizer is selected because the good results, reported in literature, that its incorporation involves. The MCM-41 mesoporous silica has been chosen since it shows similar morphological characteristics [21] to those found in SBA-15, although its hexagonal arrangement displays lower pore size due to the template required for its preparation. Furthermore, its synthesis is also less expensive than that for the SBA-15 manufacture. Binary materials constituted, on one hand, by PLA and different contents of ATBC and, on the other hand, by PLA and a given amount of MCM-41 have been also prepared additionally to the ternary PLA/plasticizer/MCM-41 composites, in order to study the existence or lack of a synergistic effect of the two additives. A thorough analysis has been performed by differential scanning calorimetry (DSC) under isothermal conditions, from either the molten or the glassy states, at different temperatures to fully understand the role that both additives exerts on PLA crystallization.

2. Materials and Methods

A commercially available polylactide (PLA) from NatureWorks® (labeled as Ingeo™ Biopolymer 6202D, with a density of 1.24 g/cm^3, and a content in *L*-isomer units of about 98 mol%) is used in this study. Its weight-average molecular weight (M_w) and molecular weight dispersity are 118,600 g/mol and 1.6, respectively, as determined from gel permeation chromatography (GPC) [22].

The ATBC (CAS 77-90-7) plasticizer was purchased from Sigma–Aldrich (Madrid, Spain) with purity equal or greater than 98.0%. Its chemical structure is shown in Scheme 1. It has a molecular weight of 402.5 g/mol and a boiling temperature of 331 °C.

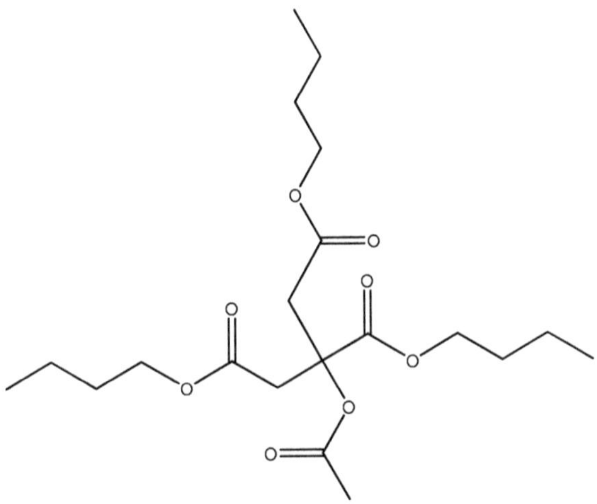

Scheme 1. Chemical structure of the ATBC plasticizer.

The MCM-41 particles were purchased from Sigma-Aldrich (specific surface area, S_{BET} = 966 m^2/g; average mesopore diameter, D_p = 2.9 nm [23]) and were used as received.

PLA, PLA-ATBC binary blends at different plasticizer contents, binary composite of PLA and MCM-41, along with PLA/ATBC/MCM-41 ternary composites were prepared through melt processing in a micro-conical twin screw Thermo Scientific Haake Minilab miniextruder operating at 190 °C, with a screw speed of 100 rpm, using a mixing time of 4 min after the full feeding. Sample labeling and nominal composition are detailed in Table 1. PLA and MCM-41 were dried previous extrusion. The former was initially placed in an oven at 85 °C for 20 min followed by a drying under vacuum at 85 °C for 2 h. The MCM-41 silica was dried under vacuum at 100 °C for 24 h.

All of these materials were subsequently processed by compression molding in a hot-plate Collin press. Initially, the material was maintained at a temperature of 190 °C without pressure for 2 min and, then, a pressure of 25 bar was applied for 5 min. Afterward, a cooling process at the relatively rapid rate of around 80 °C/min and at a pressure of 25 bar was applied to the different composites from their molten state to room temperature. These original compression-molded films were totally amorphous, as shown below from the zero neat enthalpy involved in the first melting DSC curves.

Table 1. Specimen labels and nominal composition for the different materials under study.

Sample	Nominal Material Composition (wt.%)		
	PLA	**MCM-41**	**ATBC**
PLA	100	0	0
PLAM0A5	95	0	5
PLAM0A10	90	0	10
PLAM0A20	80	0	20
PLAM5	95	5	0
PLAM5A5	90	5	5
PLAM5A10	85	5	10
PLAM5A20	75	5	20

Morphological details of the mesoporous MCM-41 were obtained by transmission electron microscopy (TEM). Measurements were performed at room temperature in a 200 kV JEM-2100 JEOL microscope. The particles were dispersed in acetone in an ultrasonic bath for 5 min and then deposited in a holder prior to observation.

Fracture surface in different sections of the films was evaluated by scanning electron microscopy (SEM) using a Philips XL30 microscope The samples were coated with a layer of 80:20 Au/Pd alloy and deposited in a holder before visualization.

Thermogravimetric analysis (TGA) was performed in a Q500 equipment of TA Instruments, under air atmosphere at a heating rate of 10 °C/min. Maximum degradation temperatures of the distinct materials are determined, as well as the actual amount of the two additives incorporated into the extruded materials prepared by extrusion.

Wide angle X-ray Diffraction (WAXD) patterns were recorded at room temperature in the reflection mode, by using a Bruker D8 Advance diffractometer provided with a PSD Vantec detector. Cu Kα radiation (λ = 0.15418 nm) was used, operating at 40 kV and 40 mA. The parallel beam optics was adjusted by a parabolic Göbel mirror with horizontal grazing incidence Soller slit of 0.12° and LiF monochromator. The equipment was calibrated with different standards. A step scanning mode was employed for the detector. The diffraction scans were collected with a 2θ step of 0.024° and 0.2 s per step.

Calorimetric analyses were carried out in a TA Instruments Q100 calorimeter connected to a cooling system and calibrated with different standards. The sample weights were around 3 mg. A temperature interval from −30 to 190 °C was studied at a heating rate of 10 °C/min in dynamic experiments. Isothermal measurements have been also performed either from the molten or the glassy state.

Details for isothermal experiments from the melt are the following: a rapid cooling, at 60 °C/min, was applied from 190 °C to the desired crystallization temperature, T_c, which was isothermally maintained for the required time. Subsequently, the heating curve at 10 °C/min is registered. The crystallization temperatures where extended to those values where the crystallization is complete after a reasonable time (total crystallization times requiring longer than around 100 min show already a very bad signal-to-noise ratio).

Details for isothermal experiments from the glassy state are the next: starting again at 190 °C, a rapid cooling is carried out at 60 °C/min, now until below the glass transition of the sample. Then, the desired crystallization temperature is reached by heating also at 60 °C/min, being isothermally maintained for the required time. Afterward, the subsequent heating curve is registered at 10 °C/min. The crystallization temperatures where extended up to those temperatures before the cold crystallization is observed in the heating ramp for reaching the crystallization temperature.

All the DSC curves presented throughout this work are actual curves without normalization for the real content of PLA in the sample. That normalization has been performed, however, for the enthalpies used to estimate crystallinity degree of the different sample experiments, i.e., the values achieved are then the normalized crystallinity. For the determination of the crystallinity, 93.1 J/g was used as the enthalpy of fusion of a perfectly crystalline material [24,25].

3. Results and Discussion

Figure 1a shows the ordered hexagonal arrangements that constitute the interior of the particles of the mesoporous MCM-41 silica. This inner part consists in a well-defined hollow channel structure with pore diameters of around 3 nm, where PLA chain segments and/or molecules of plasticizer can be embedded. Experimental conditions used during extrusion turn out appropriate, as deduced from the SEM fracture picture found in the PLAM5 composite depicted in Figure 1b, where a homogeneous distribution of silica within the matrix of PLA is exhibited together with the absence of inorganic domains of large size.

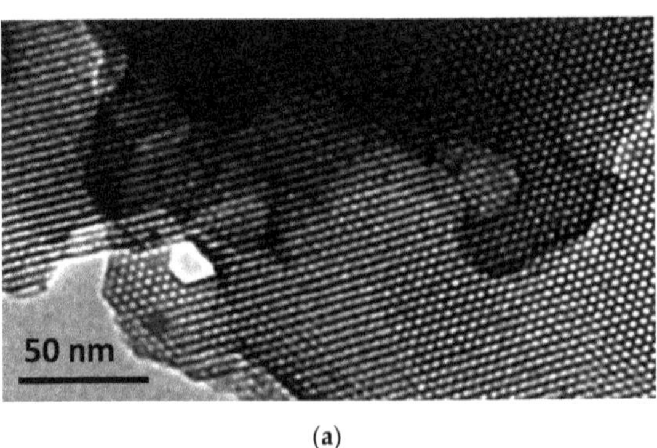

(a)

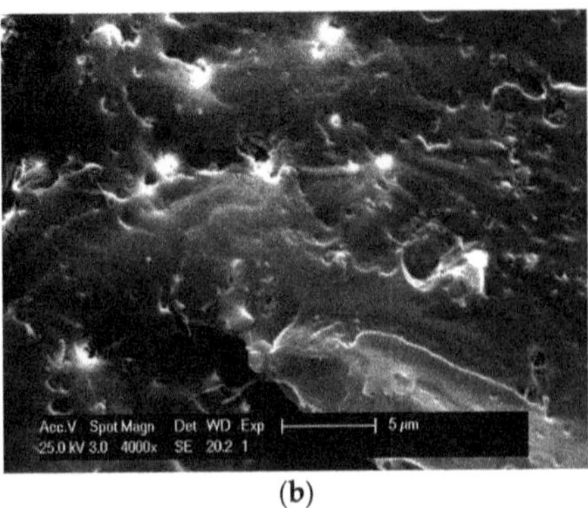

(b)

Figure 1. (**a**) TEM micrographs of MCM-41. (**b**) SEM fracture picture for the PLAM5 composite.

Thermogravimetry measurements are carried out in order to learn how presence of: (a) individual plasticizer in the PLA-ATBC blends, (b) MCM-41 in binary composites or (c) both additives in plasticized and filled PLA ternary materials, affects the thermal stability of this polymeric matrix. Furthermore, these TGA experiments allow estimating the actual material content for the different constituents (PLA, ATBC plasticizer and MCM-41 silica) in the several composites. Figure 2 shows the TGA curves under air atmosphere for the different samples, indicating that PLA maximum degradation temperature is shifted to higher temperatures (see Table 2) by incorporation of ATBC plasticizer in the binary blends.

Table 2. Results derived from TGA experiments performed under air: temperature for 1% degradation ($T_{1\%}$), the maximum degradation temperature (T_{max}^{PLA}) and the actual material content based on TGA analysis for the different constituents: PLA, ATBC plasticizer and MCM-41 silica.

Specimen	$T_{1\%}$ (°C)	T_{max}^{PLA} (°C)	TGA Content (wt.%)		
			PLA	MCM-41	ATBC
PLA	221.0	315.5	100	0	0
PLAM0A5	159.0	338.0	94	0	6
PLAM0A10	154.0	341.5	91	0	9
PLAM0A20	148.0	341.5	82	0	18
PLAM5	260.0	358.5	95.3	4.7	0
PLAM5A5	161.0	337.0	90.1	3.9	6
PLAM5A10	149.0	358.0	83.7	5.3	11
PLAM5A20	145.0	344.0	77.2	4.8	18

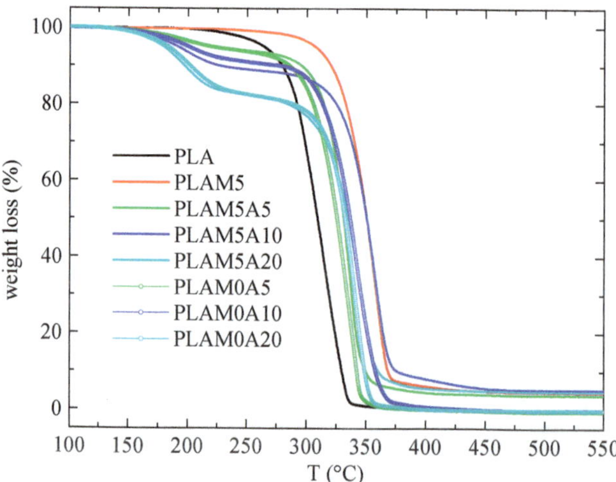

Figure 2. TGA curves under air atmosphere for the different materials analyzed.

At the lowest amount of ATBC, the increase is around 23 °C. This rise is even slightly larger for the two higher contents of ATBC used. A much important displacement is ob-served in the PLAM5 specimen, which is the binary composite of PLA and MCM-41 silica. Thus, inclusion of mesoporous MCM-41 into PLA by extrusion leads to a very remarkable improvement of PLA thermal stability. The shift of the maximum degradation temperature (T_{max}^{PLA}) to higher temperatures has been also previously described when adding talc to PLA [26] and in PLA composites incorporating modified clays [27]. The combined effect of ATBC plasticizer and MCM-41 silica leads to an increase of the PLA thermal stability compared with that exhibited by the neat polymeric matrix, showing an optimized T_{max}^{PLA} value in the sample PLAM5A10. This type of behavior of a maximum effect at intermediate plasticizer content has been already reported [4]. Nevertheless, a straightforward variation with the amount of the two additives is not noticed. Thus, the values found in PLAM5A5 and PLAM5A20 are analogous to those observed in PLAM0A5 and PLAM0A20, respectively, while that shown by PLAM5A10 is similar to that noted in PLAM5. A full understanding of the thermal stability of these ternary systems requires a more complete analysis, which is out of the scope of this investigation.

Table 2 also displays that the $T_{1\%}$ in air is considerably smaller in the materials that contain the ATBC additive, since this plasticizer is the first compound that is lost through evaporation. Moreover, the actual content of the different constituents in the several materials are also collected in Table 2.

As commented in the Introduction, ternary systems based on PLA as a matrix together with both a plasticizer and a filler, as minor constituents, are not commonly analyzed. Therefore, an exhaustive analysis of the crystallization capability exhibited by these materials is mandatory before the evaluation of any type of properties. Figure 3 shows the DSC curves obtained for the first heating process in all of the examined samples. Several phase transitions are noticeable: glass transition, cold crystallization and melting processes in order of increasing temperatures. It is important to mention that the neat enthalpy involved in the entire DSC curves is negligible in all cases, indicating the absence of crystallinity in the initial extruded samples. That means that these specimens rapidly cooled from the melt show a complete amorphous nature.

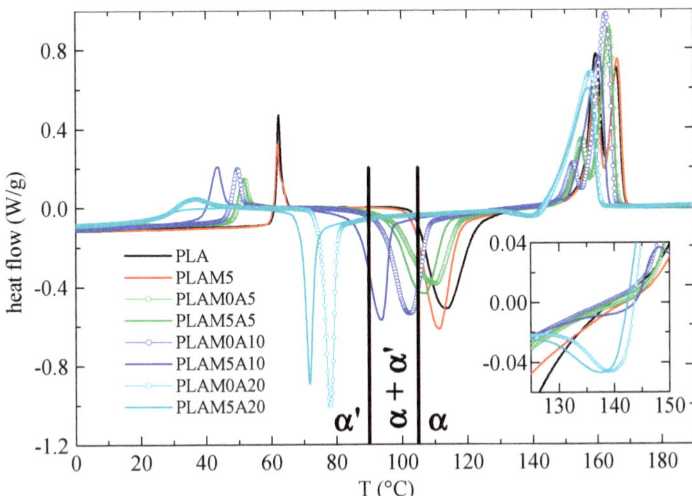

Figure 3. DSC first melting curves (endo up) at 10 °C/min of the different compression-molded materials. The vertical lines are the corresponding values reported in reference [22], showing the temperature intervals where the pure α' or α modifications are expected to be observed (or the interval where both forms coexist) in this PLA under analysis.

Location of the glass transition moves very remarkably to lower temperatures as the ATBC content increases in the samples. In fact, T_g depression is around 34 °C, as seen in Figure 3 and quantified in Figure 4a. This reduction is similar to that previously reported in literature [4,28] and it is ascribed to the good compatibility of both PLA and ATBC. The position of T_g for the neat ATBC is about −81 °C [28] and, thus, a progressive decrease of the T_g in the PLA-ATBC blends is expected as the ATBC content increases if ATBC segregation does not take place.

Furthermore, the different curves show a rather prominent aging peak overlapped at the top of the glass transition. Intensity of this aging peak decreases with increasing ATBC content in the sample, being significantly small for sample PLAM0A20, and negligible for PLAM5A20. One of the reasons for this behavior is that the glass transition of these samples has been moved noticeably to a lower location compared with that for the neat PLA, taking place now at about room temperature (around 25–30 °C). Structural recovery is much faster [29] in these two specimens.

The cold crystallization (T_{cc}^{F1}) appears at considerably lower temperatures as the ATBC amount rises regarding that shown in the pristine polymeric matrix, meaning that PLA is able to crystallize more rapidly induced by presence of plasticizer. An additional favorable contribution is observed by incorporation of the MCM-41 particles due to their nucleant effect, so that the sample with the smallest T_{cc}^{F1} is PLAM5A20 (see Figure 4b). Thus, this nucleating influence is noticed when pure PLA is compared with PLAM5 and when comparison is established between the other plasticized blends with their respective ternary composites, i.e., PLAM0A5 with PLAM5A5 and PLAM0A10 with PLAM5A10.

Reason behind this behavior is believed to be associated with the combined action of incorporation of ATBC and MCM-41 particles within the PLA matrix. On one hand, the MCM-41 silica acts as nucleating agent, similar to the effect observed by addition of mesoporous SBA-15 into PLA [17,18]. This feature allows enhancing crystal nucleation rate due to the reduction of the nucleation energy barrier of PLA via absorption of PLA chains on external and internal surface of MCM-41 [30,31] through the silanol groups existing in this silica [32]. Nevertheless, the crystal growth rate could be constrained by the PLA mobility. Therefore, the T_{cc}^{F1} in PLAM5 is expected to decrease due to the addition of

MCM-41, as Figure 3 shows. On the other hand, the ATBC improves the mobility of PLA chains by plasticization, as confirmed by the shift of T_g values to lower temperatures with increasing ATBC content. Accordingly, the crystal growth rate of PLA should be enhanced when adding MCM-41 and ATBC, turning out in the decrease of the T_{cc}^{F1} of PLA, as seen in Figures 3 and 4b.

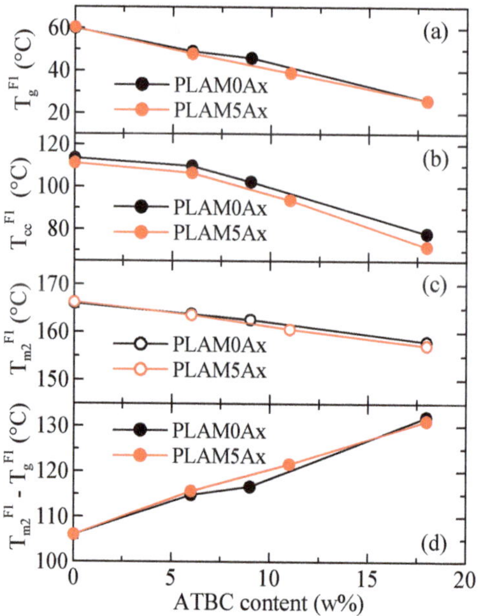

Figure 4. Variation with the ATBC content for the two series of samples of: (**a**) the glass transition temperature; (**b**) the cold crystallization; (**c**) the second melting temperature; (**d**) and the difference between the second melting temperature and the glass transition. All of these magnitudes have been determined in the first heating run of the samples.

Moreover, the enthalpy involved in the cold crystallization is also sensitive to the presence of MCM-41 and ATBC. Thus, the values deduced from Figure 3, after normalization to the actual PLA content at each sample, are the following: 38.1, 39.6, 41.9 and 43.8 J/g for the PLA, PLAM0A5, PLAM0A10 and PLAM0A20 specimens, respectively, and 39.2, 40.4, 42.7 and 44.6 J/g for the PLAM5, PLAM5A5, PLAM5A10 and PLAM5A20 materials, respectively. Two conclusions are derived from these values: first, in the two series of samples, PLAM0Ax and PLAM5Ax, the enthalpy increases with the ATBC content, and, second, a further increase is also noted when the MCM-41 is present, so that enthalpies of cold crystallization are in the PLAM5Ax composites higher than those in the PLAM0Ax samples.

On the other hand, according to previous findings [22] reported for this specific PLA, represented in Figure 3 by the vertical lines, the pure α' modification is expected to be developed in samples PLAM0A20 and PLAM5A20, while the pure α form should be obtained for PLA and PLAM5, with mostly a mixture of the two modifications for the remaining samples. Those findings reported in literature for this specific PLA (with a content in D-isomer of about 2 mol%) were based on the dependence on crystallization temperatures of the location for the main reflections derived from X-ray diffraction measurements. The variation in the spacing position of the more intense diffractions was then related to the α' to α transition [22]. Existence of only the α' lattice, or of a large amount of this polymorph, can be also characterized by DSC through appearance of the transformation from the α' to the α crystalline form during the heating process by means of a small exotherm prior the dominant melting process [24,25,33,34]. This transformation is only well noted in PLAM0A20 and PLAM5A20 (at around 130–140 °C), and somewhat in PLAM5A10, as shown in the inset of Figure 3. This observation is in agreement with the statements aforementioned about the expected modification to be obtained.

Figures 3 and 4c show that location of the maximum of melting process also takes place at lower temperatures in these either only plasticized or plasticized and filled materials. The depression in T_m, of around 9 °C as seen in Figure 4c, is much smaller than the one for T_g. As a consequence, interval for the actual crystallization occurring at "normal" rates increases considerably on passing from the pristine PLA to PLAM0A20 or PLAM5A20, as noticed in Figure 4d.

The greater crystallization rate deduced from the results in Figure 3 for the samples plasticized with ATBC and/or filled with MCM-41 is also noted on the cooling curves from the melt represented in Figure 5a. Thus, the neat PLA sample does not show any crystallization at the cooling rate of 10 °C/min, contrary to the other samples, where the exotherm increases in intensity, especially in those materials containing both ATBC and MCM-41. This is also clearly deduced from Figure 5b, where the DSC crystallinity (normalized to the actual PLA content in the material) is represented as function of the ATBC content. A considerable increase in crystallinity is observed for the PLAM0Ax samples, which is even higher for the PLAM5Ax specimens, and, importantly, the difference between the two series is increasingly greater. It is deduced, therefore, that the simultaneous presence of ATBC and MCM-41 leads to an especially noticeable increase of the crystallization rate of PLA. It seems that there is a synergistic effect of mesoporous MCM-41 and ATBC plasticizer on the nucleation ability of PLA, leading not only to a higher crystallization rate, but also extending its crystallization capability.

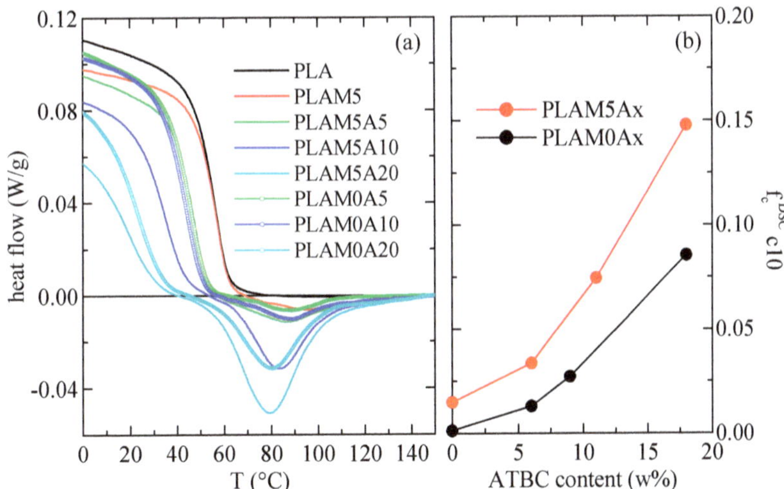

Figure 5. DSC cooling curves (exo down) at 10 °C/min of: (**a**) the different samples and (**b**) DSC crystallinity on cooling at that rate as function of the ATBC content in the sample.

Several experiments of isothermal crystallization have been performed in the different samples by DSC either from the glassy state or from the melt in order to ascertain the increase in crystallization rate. As an example, Figure 6a shows the crystallization isotherms, both from the glassy state or from the melt, and the subsequent heating curves at 10 °C/min (Figure 6b) for sample PLAM5A10.

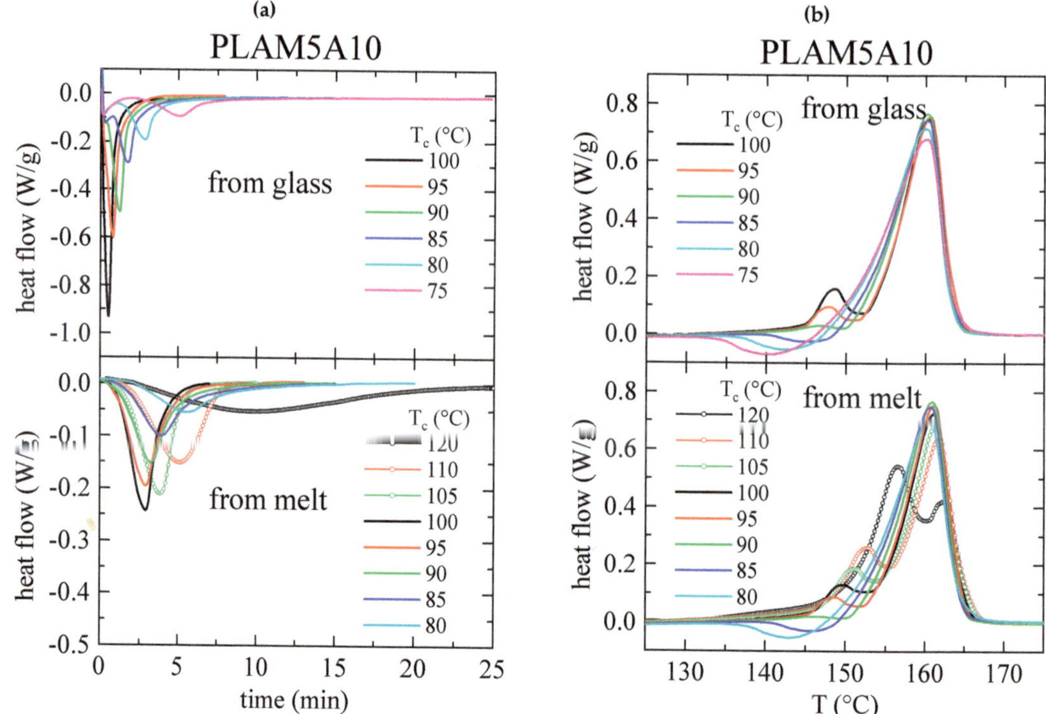

Figure 6. (**a**) Crystallization isotherms (exo down) for sample PLAM5A10, either from the glassy state or from the melt, and (**b**) subsequent heating curves (endo up), at 10 °C/min.

Important differences are deduced between the isotherms attained from the glass or from melt. Crystallization rate is much faster when it is initiated from the glassy state, as expected, and it increases continuously with the crystallization temperature, in such a way that it is not possible to observe the isotherm above 100 °C, since the crystallization is so fast that it begins during the heating, at 60 °C/min, to reach the isothermal crystallization temperature.

On the contrary, the isotherms from the melt show a crystallization rate considerably much smaller, and the crystallization can be performed up to 120 °C in a reasonable time. Nevertheless, this much lower rate makes the crystallization range in the region of low T_cs shorter than when crystallization starts from the glass. Moreover, now the crystallization rate shows a maximum at around 100 °C, as expected from the well-known transport and free-energy terms in the crystallization rate [35,36]. A more detailed study of those crystallization rates is made below, extended also to all samples analyzed.

Regarding the subsequent heating, the melting curves for a particular T_c are rather similar for both crystallizations, from the glass or from the melt, as observed in Figure 6b. In both cases, the transformation from the α' to α modification, which is characterized by a melting-recrystallization process [24,25,33,34], is clearly observed until an isothermal T_c of around 90 °C. Above that temperature, the melting curves show a bimodal melting endotherm, the low temperature component, termed as T_{m1}, supposedly arising from the melting of the real imperfect crystals initially formed at the crystallization temperature, while the second component, T_{m2}, is due to the final melting of the recrystallized imperfect crystals [22]. It is interesting to note that the upper limit for observing the transition from the α' form to the α modification overlaps with the beginning of the observation T_{m1}, so that separation between these two processes (the transition α' to α and T_{m1}) at the T_c of

90 °C is difficult to be resolved. Again, a more detailed study of these aspects is made below for the different samples.

Thus, crystallization rate has been estimated from the inverse of time for the 50% transformation in the isotherms, i.e., the half crystallization time, $t_{0.5}$. The values of $1/t_{0.5}$ are represented in Figure 7 for the different samples and the two modes of crystallization. Crystallization rate is much faster for all the samples when crystallizing from the glass (note that the Y scale in the data from melt has been amplified 5 times in Figure 7b), as mentioned above for PLAM5A10 and as expected. It is observed that, when isothermal crystallization occurs from the glass, the upper limit of T_c for reasonable crystallization times occurs at progressively inferior temperatures as the ATBC + MCM41 content increases in the sample (according with the decrease in the cold crystallization temperature obtained in dynamic experiments represented in Figures 3 and 4). And, therefore, the maximum in the crystallization rate is not observed in these specimens crystallized from the glass, since the crystallization is so fast that it begins, as mentioned, during the heating, at 60 °C/min, before reaching the isothermal crystallization temperature.

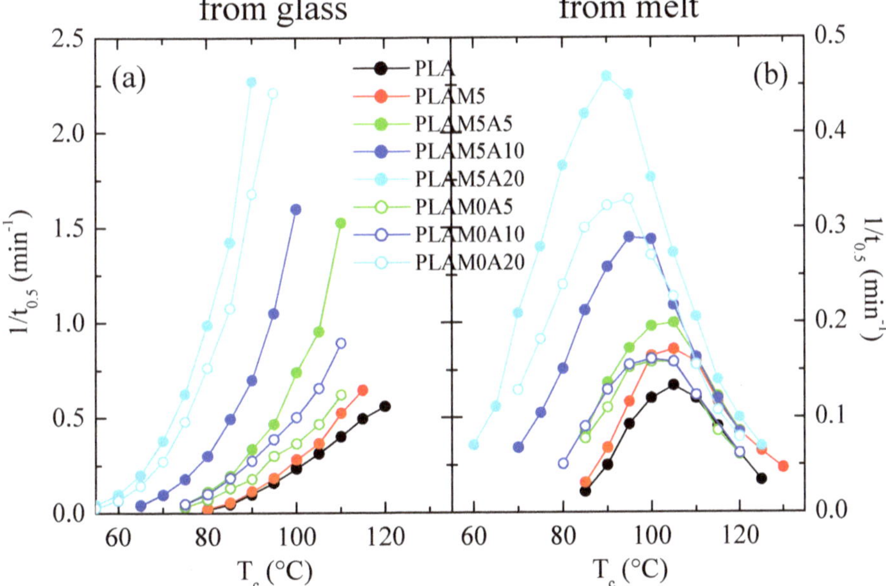

Figure 7. Variation with T_c of the inverse of the half crystallization time in the isotherms for the different samples, for isothermal crystallization: (**a**) from the glass and (**b**) from the melt. Note that the Y scale in the data from melt has been amplified 5 times.

A maximum is, however, noticed for the crystallizations from the melt, as deduced from Figure 7b. This maximum arises from the well-known fact that there are two factors, the free-energy term and the transport one, which determine the crystallization rate of polymers [35,36], as mentioned above. And, these two terms show opposite temperature coefficients. It can be observed that the maximum increases in absolute rate and occurs at progressively lower temperatures as the content in ATBC and MCM-41 is raised. In fact, the coordinates of the maxima, i.e., $1/t_{0.5}$ and T_c^{max}, are represented in Figure 8 as a function of the ATBC content in the material.

The subsequent melting curves show rather interesting features. Concerning the curves of materials isothermally crystallized at $T_c = 85\,°C$ (Figure 10a), the α' to α transition occurs for samples PLA and PLAM5 at higher temperatures than for the other specimens. Moreover, this transformation is more clearly noticeable (the apparent heat flow change is higher) in those two samples, while the heat flow change is progressively decreasing in such a way that is rather small (but appreciable) for sample PLAM5A20. It is true, however, that the DSC curves in Figure 10 are the actual curves, not normalized to the real PLA content in the sample. Furthermore, there is continuous decrease of the main peak temperature on passing from PLA to PLAM5A20 (see below the corresponding values), in the melting endotherms.

Regarding the results for $T_c = 120\,°C$, several features are deduced from Figure 10b. Firstly, the two constituents of the now bimodal melting endotherms also decrease in temperature from PLA to PLAM5A20 (see below). Secondly, the recrystallized component (T_{m2}) is increasing considerably its intensity in relation to T_{m1} as the content in ATBC increases and MCM-41 is added. And, finally, the two melting peaks are getting wider as that content is raised. It seems, therefore, that the presence of ATBC leads to a wider distribution in the size of the crystallites.

The results for all the materials and crystallization temperatures regarding the α' to α transition temperature together with the melting temperatures are represented in Figures 11 and 12, respectively. The temperatures for the transition from the α' to α modification, displayed in Figure 11, indicate, first, that rather similar values are obtained when crystallizing from the glass or from the melt. Moreover, a small, but appreciable, decrease of those temperatures is observed as the ATBC content increases, extending also continuously the interval of actual crystallization temperatures.

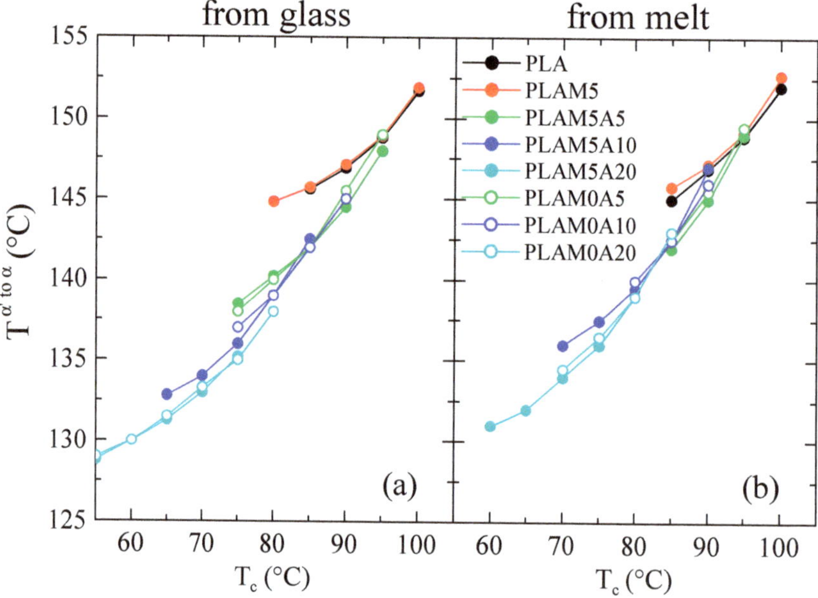

Figure 11. Variation with T_c of the temperature for transition from the α' to α for the different samples, when crystallizing (**a**) from the glass, and (**b**) from the melt.

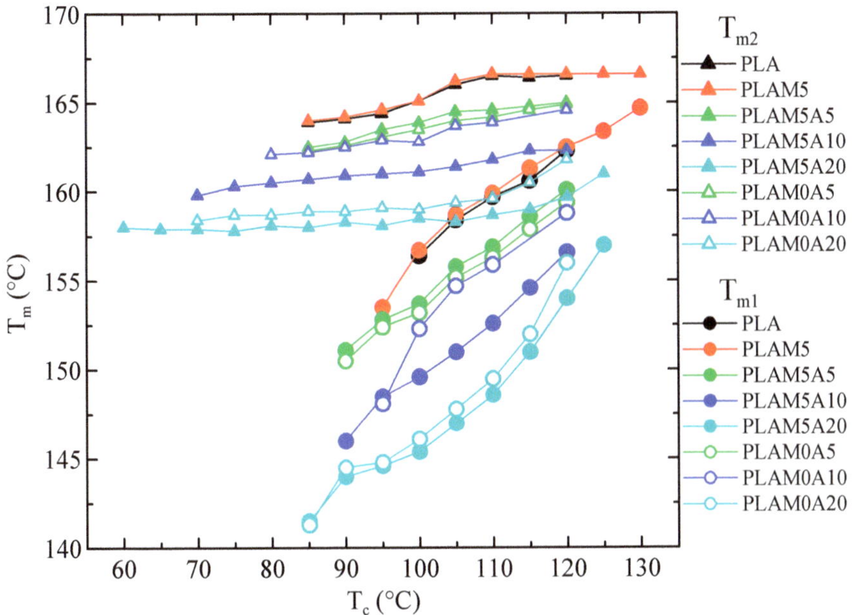

Figure 12. Variation with T_c of the melting temperatures for the different samples, when crystallizing from the melt: T_{m1} (circles) and T_{m2} (triangles).

Regarding the melting temperatures (Figure 12), the values of T_{m1} for the low T_c's appear to decrease abnormally, since they are overlapped to the transition from the α' form to the α polymorph, as commented. For higher T_c's, however, a continuous increase in the values of T_{m1} is observed for all materials, in such a way that they tend to those of T_{m2}, where the variation with T_c is much smaller. Moreover, and comparing the different samples, the presence of ATBC and MCM-41 leads to significantly smaller values for the two melting temperatures (although this decrease is not as high as the one for T_g, as seen in Figure 4).

The values of the PLA crystallinity developed in the different materials and crystallization conditions have been estimated from the enthalpies involved in the corresponding melting curves and after normalization to the actual PLA content in the sample. For instance, Figure 13a displays the values of the so determined DSC crystallinity as a function of the crystallization temperature in the experiments conducted from the melt. It can be observed that there is a more or less continuous increase of the crystallinity with the ATBC + MCM content in the sample. A more clear picture is deduced from Figure 13b, representing the variation with the ATBC content of the degrees of crystallinity after isothermal crystallization at 85 and 120 °C for the two series of samples. It is evident, firstly, that the presence of ATBC leads to an increase of the crystallinity of around 0.02–0.03 units at the maximum ATBC content compared with the neat PLA. Moreover, there is an additional increase of 0.01–0.02 units more when mesoporous MCM-41 particles are also present. Again, the simultaneous presence of ATBC and MCM-41 involves a beneficial effect, now in the crystallinity values, in these materials.

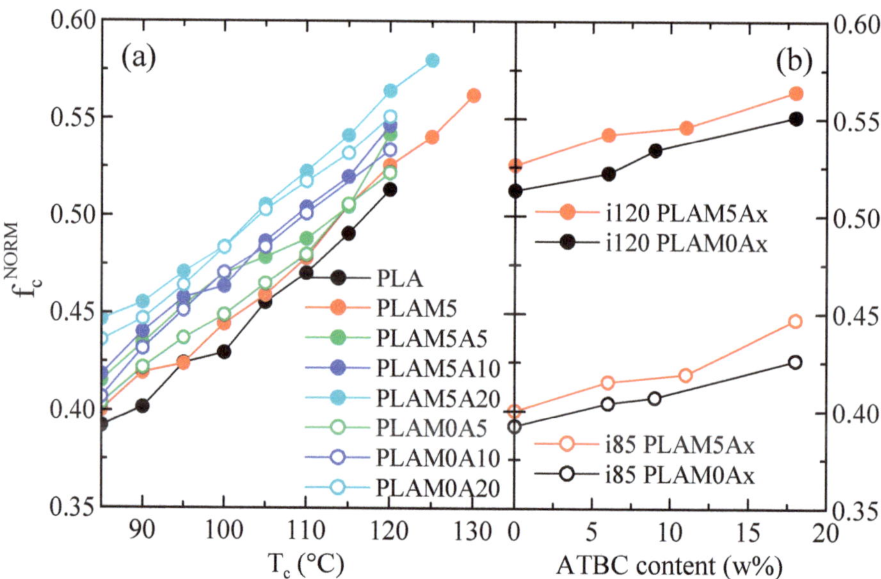

Figure 13. Variation of the DSC crystallinity, normalized to the actual PLA content, for the experiments conducted from the melt, as a function of T_c for the different samples (**a**), and of the ATBC content after isothermal crystallization at 85 and 120 °C for the two series of samples (**b**).

As a final aspect, some preliminary X-ray diffraction experiments have been performed in samples isothermally crystallized in the DSC and brought to room temperature at the end of the isotherm. Two representative materials (neat PLA and PLAM5A10) have been crystallized from the glass at two temperatures, and the corresponding X-ray diffractograms are shown in Figure 14. It is observed that the diffraction for the sample of PLA cold crystallized at 85 °C leads to a (110/200) diffraction peak centered very close to that reported for the pure α' modification [33], while this reflection in the specimen isothermally crystallized at 120 °C is close to the characteristic for the α modification. Similar features apply for the (203) diffraction, shown in the inset of Figure 14. Concerning the PLAM5A10 material cold crystallized at 70 and 120 °C, these profiles are close to the α' and α forms, respectively, but not as close as the PLA samples. Moreover, the PLAM5A10 profiles display a somewhat higher width. The reason for this behavior may be related to the fact that the DSC pans for performing these X-ray measurements were not hermetically close in order to facilitate the specimens removing prior X-ray experiments. That may lead to some uncertainty in the real temperatures inside the sample. Experiments are being planned to study the crystallization of the different samples under real-time variable-temperature X-ray conditions using synchrotron radiation, and thus analyze the possible differences of the crystalline modifications between PLA and the materials with ATBC and MCM-41. Furthermore, a complete evaluation of several mechanical properties is under progress in order to analyze and understand the influence of presence of the selected plasticizer and nucleant agent in the final performance of these, either fully amorphous or isothermally crystallized, plasticized and/or filled materials.

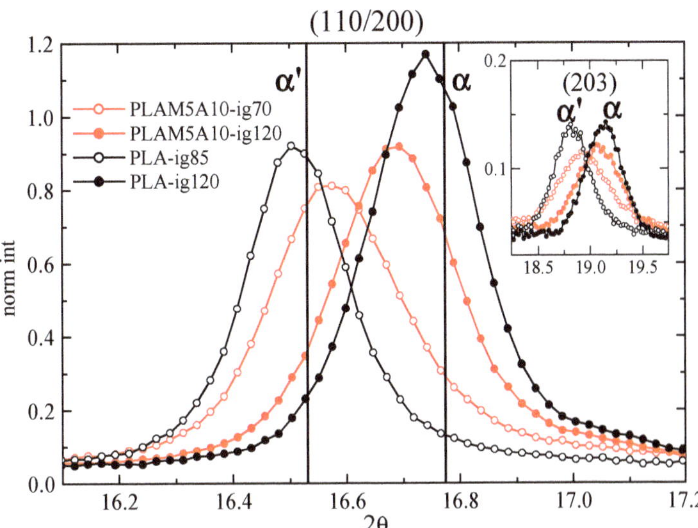

Figure 14. Room-temperature X-ray diffractograms of different samples after cold crystallization from the amorphous glassy state at the indicated temperatures, showing the angular interval of the main (110/200) diffraction and that for the (203) reflection one in the inset. The vertical lines are the corresponding values reported in ref. [33] for the pure α' and α modifications.

4. Conclusions

Binary blends and dual compounds, the former based on an *L*-rich PLA and ATBC as plasticizer and the latter on PLA and MCM-41 as nucleating agent, together with ternary composites consisting of PLA, ATBC at several compositions and MCM-41 silica at a given content, were processed by melt extrusion. An appropriate distribution of mesoporous particles was found in the different loaded materials along with the absence of bulky silica domains.

The combined effect of ATBC and MCM-41 silica leads to an increase of the PLA thermal stability, showing the optimized situation for the sample PLAM5A10

Several phase transitions are noticeable by DSC during the first heating process under dynamic conditions in all of the materials: glass transition, cold crystallization and melting processes in order of increasing temperatures. Location of the glass transition moves very remarkably to lower temperatures as the ATBC content increases in the samples. Motion capability of amorphous phase is not, however, affected at a given plasticizer amount by the incorporation of MCM-41 in the ternary systems.

The cold crystallization is also considerably shifted to lower temperatures as the ATBC amount rises, in relation to that shown in the pristine matrix. Thus, PLA is able to crystallize more rapidly boosted by presence of plasticizer. Furthermore, an additional favorable contribution is provided by inclusion of the MCM-41 particles in the ternary composites. Position of the maximum of melting process also takes place at lower temperatures in these either only plasticized or plasticized and loaded systems, this decrease being, however, much smaller than the one for T_g. As a consequence, the crystallization window is considerably enlarged by the presence of ATBC and MCM-41.

A higher crystallization rate is deduced by addition of the plasticizer and/or mesoporous silica under dynamic cooling at 10 °C/min. The neat PLA does not show any crystallization at that cooling rate, contrary to the other materials. Thus, a considerable increase in crystallinity is observed for the PLAM0Ax samples, which is even higher for the PLAM5Ax specimens.

PLA crystallization rate in isothermal experiments is much faster from the glassy state than from the melt, as expected. A maximum rate is found at around 90–105 °C in the experiments conducted from the melt, showing a very important rise of the crystallization rate as the content in ATBC and MCM-41 increases. That maximum in the rate is, however, not observed in the specimens crystallized from the glassy state, since the crystallization is so fast that it begins during the heating before reaching the isothermal crystallization temperature.

Summarizing, a synergistic effect of mesoporous MCM-41 silica and ATBC plasticizer seems to occur on the nucleation ability of PLA, thus leading to considerably higher crystallization rates and broadening the crystallization window of this PLA.

Author Contributions: Conceptualization, E.B.-B., M.L.C. and E.P.; methodology, E.B.-B., R.B.-G., M.L.C. and E.P.; software, E.P.; formal analysis, E.P. and E.B.-B.; investigation, E.B.-B., R.B.-G., T.M.D.-R., M.L.C. and E.P.; resources, E.P. and M.L.C.; writing—original draft preparation, E.P. and M.L.C.; writing—review and editing, E.B.-B., R.B.-G., T.M.D.-R., M.L.C. and E.P.; supervision, M.L.C. and E.P.; project administration, E.P. and M.L.C.; funding acquisition, E.P. and M.L.C. All authors have read and agreed to the published version of the manuscript.

Funding: This research was funded by MCIN/AEI/10.13039/501100011033, grant number PID2020-114930GB-I00. Grant BES-2017-082284 was also funded by MCIN/AEI/10.13039/501100011033 and, thus, TMDR is also thankful for this pre-doctoral funding.

Data Availability Statement: Data are contained within the article.

Acknowledgments: Authors are grateful to the Characterization Service for SEM, TGA, and X-ray Diffraction facilities and to the Service of Physicochemical Characterization of Polymers for DSC experiments, both services at ICTP-CSIC. Authors are also grateful to the personnel of these two Services for the support.

Conflicts of Interest: The authors declare no conflict of interest. The funders had no role in the design of the study; in the collection, analyses, or interpretation of data; in the writing of the manuscript; or in the decision to publish the results.

References

1. Celebi, H.; Gunes, E. Combined effect of a plasticizer and carvacrol and thymol on the mechanical, thermal, morphological properties of poly(lactic acid). *J. Appl. Polym. Sci.* **2018**, *135*, 45895. [CrossRef]
2. Scatto, M.; Salmini, E.; Castiello, S.; Coltelli, M.-B.; Conzatti, L.; Stagnaro, P.; Andreotti, L.; Bronco, S. Plasticized and Nanofilled Poly(lactic acid)-Based Cast Films: Effect of Plasticizer and Organoclay on Processability and Final Properties. *J. Appl. Polym. Sci.* **2013**, *127*, 4947–4956. [CrossRef]
3. Loudin, D.; Bizot, H.; Colonna, P. "Antiplasticization" in starch-glycerol films? *J. Appl. Polym. Sci.* **1997**, *63*, 1047–1053. [CrossRef]
4. Maiza, M.; Benaniba, M.T.; Massardier-Nageotte, V. Plasticizing effects of citrate esters on properties of poly(lactic acid). *J. Polym. Eng.* **2016**, *36*, 371–380. [CrossRef]
5. Ljungberg, N.; Wesslén, B. Preparation and properties of plasticized poly(lactide) films. *Biomacromolecules* **2005**, *6*, 1789–1796. [CrossRef] [PubMed]
6. Jacobsen, S.; Fritz, H.G. Plasticizing poly(lactide)-the effect of different plasticizers on the mechanical properties. *Polym. Eng. Sci.* **1999**, *39*, 1303–1310. [CrossRef]
7. Hu, Y.; Hu, Y.S.; Topolkaraev, V.; Hiltner, A.; Baer, E. Crystallization and phase separation in blends of high stereoregular poly(lactide) with poly(ethene glycol). *Polymer* **2003**, *44*, 5681–5689. [CrossRef]
8. Hu, Y.; Rogunova, M.; Topolkaraev, V.; Hiltner, A.; Baer, E. Ageing of poly(lactide)/poly(ethylene glycol) blends. Part 1.Poly(lactide) with low stereoregularity. *Polymer* **2003**, *44*, 5701–5710. [CrossRef]
9. Kulinski, Z.; Piorkowska, E.; Gadzinowska, K.; Stasiak, M. Plasticizarion of poly(lactide) with poly(propylene glycol). *Biomacromolecules* **2006**, *7*, 2128–2135. [CrossRef]
10. Focarete, M.L.; Dobrzynski, M.S.; Kowalczuk, M. Miscibility and Mechanical Properties of Blends of (l)-Lactide Copolymers with Atactic Poly(3-hydroxybutyrate). *Macromolecules* **2002**, *35*, 8472–8477. [CrossRef]
11. Blümm, E.; Owen, A.J. Miscibility, crystallization and melting of poly (3-hydroxybutyrate)/poly (L-lactide) blends. *Polymer* **1995**, *36*, 4077–4081. [CrossRef]
12. Ljungberg, N.; Andersson, T.; Wesslén, B. Film extrusion and film weldability of poly(lactide) plasticized with triacetine and tributyl citrate. *J. Appl. Polym. Sci.* **2003**, *88*, 3239–3247. [CrossRef]
13. Ljungberg, N.; Wesslén, B. Tributyl citrate oligomers as plasticizers for poly(lactide): Thermo-mechanical film properties and aging. *Polymer* **2003**, *44*, 7679–7688. [CrossRef]

14. Ren, Z.; Dong, L.; Yang, Y. Dynamic mechanical and thermal properties of plasticized poly (lactic acid). *J. Appl. Polym. Sci.* **2006**, *101*, 1583–1590. [CrossRef]
15. Saeidlou, S.; Huneault, M.A.; Li, H.; Park, C.B. Poly(lactic acid) crystallization. *Prog. Polym. Sci.* **2012**, *37*, 1657–1677. [CrossRef]
16. Petchwattana, N.; Narupai, B. Synergistic effect of talc and titanium dioxide on poly (lactic acid) crystallization: An investigation on the injection molding cycle time reduction. *J. Polym. Environ.* **2019**, *27*, 837–846. [CrossRef]
17. Díez-Rodríguez, T.M.; Blázquez-Blázquez, E.; Pérez, E.; Cerrada, M.L. Composites Based on Poly(Lactic Acid) (PLA) and SBA-15: Effect of Mesoporous Silica on Thermal Stability and on Isothermal Crystallization from Either Glass or Molten State. *Polymers* **2020**, *12*, 2743. [CrossRef]
18. Díez-Rodríguez, T.M.; Blázquez-Blázquez, E.; Martínez, J.C.; Pérez, E.; Cerrada, M.L. Composites of a PLA with SBA-15 mesoporous silica: Polymorphism and properties after isothermal cold crystallization. *Polymer* **2022**, *241*, 124515. [CrossRef]
19. Díez-Rodríguez, T.M.; Blázquez-Blázquez, E.; Barranco-García, R.; Pérez, E.; Cerrada, M.L. Synergistic effect of mesoporous SBA-15 particles and processing strategy for improving PLA crystallization capability in their composites. *Macromol. Mater. Eng.* **2022**, *307*, 2200308. [CrossRef]
20. Zhao, D.Y.; Feng, J.L.; Huo, Q.S.; Melosh, N.; Fredrickson, G.H.; Chmelka, B.F.; Stucky, G.D. Triblock copolymer syntheses of mesoporous silica with periodic 50 to 300 angstrom pores. *Science* **1998**, *279*, 548–552. [CrossRef]
21. Beck, J.S.; Vartuli, J.C.; Roth, W.J.; Leonowicz, M.E.; Kresge, C.T.; Schmitt, K.D.; Chu, C.T.-W.; Olson, D.H.; Sheppard, E.W.; McCullen, S.B.; et al. A new family of mesoporous molecular sieves prepared with liquid crystal templates. *J. Am. Chem. Soc.* **1992**, *114*, 10834. [CrossRef]
22. Díez-Rodríguez, T.M.; Blázquez-Blázquez, E.; Pérez, E.; Cerrada, M.L. Influence of content in D isomer and incorporation of SBA-15 silica on the crystallization ability and mechanical properties in PLLA based materials. *Polymers* **2022**, *14*, 1237. [CrossRef] [PubMed]
23. Watanabe, R.; Hagihara, H.; Sato, H. Structure-property relationship of polypropylene-based nanocomposites by dispersing mesoporous silica in functionalized polypropylene containing hydroxyl groups. Part 1: Toughness, stiffness and transparency. *Polym. J.* **2018**, *50*, 1057–1065. [CrossRef]
24. Zhang, J.; Tashiro, K.; Tsuji, H.; Domb, A.J. Disorder-to-order phase transition and multiple melting behavior of poly(l-lactide) investigated by simultaneous measurements of WAXD and DSC. *Macromolecules* **2008**, *41*, 1352–1357. [CrossRef]
25. Beltrán, F.R.; de la Orden, M.U.; Lorenzo, V.; Pérez, E.; Cerrada, M.L.; Martínez-Urreaga, J. Water-induced structural changes in poly(lactic acid) and PLLA-clay nanocomposites. *Polymer* **2016**, *107*, 211–222. [CrossRef]
26. Lee, C.; Pang, M.M.; Koay, S.C.; Choo, H.L.; Tshai, K.Y. Talc filled polylactic-acid biobased polymer composites: Tensile, thermal and morphological properties. *SN Appl. Sci.* **2020**, *2*, 354. [CrossRef]
27. Zhou, Q.; Xanthos, M. Nanosize and microsize clay effects on the kinetics of the thermal degradation of polylactides. *Polym. Degrad. Stab.* **2009**, *94*, 327–338. [CrossRef]
28. Baiardo, M.; Frisoni, G.; Scandola, M.; Rimelen, M.; Lips, D.; Ruffieux, K.; Wintermantel, E. Thermal and Mechanical Properties of Plasticized Poly(L-lactic acid). *J. Appl. Polym. Sci.* **2003**, *90*, 1731–1738. [CrossRef]
29. Struik, L.C.E. *Physical Aging in Amorphous Polymers and Other Materials*; Elsevier: Amsterdam, The Netherlands, 1978.
30. Barranco-García, R.; Gómez-Elvira, J.M.; Ressia, J.A.; Quinzani, L.; Vallés, E.M.; Pérez, E.; Cerrada, M.L. Variation of Ultimate Properties in Extruded iPP-Mesoporous Silica Nanocomposites by Effect of iPP Confinement within the Mesostructures. *Polymers* **2020**, *12*, 70. [CrossRef]
31. Díez-Rodríguez, T.M.; Blázquez-Blázquez, E.; Antunes, N.L.C.; Ribeiro, M.R.; Pérez, E.; Cerrada, M.L. Confinement in Extruded Nanocomposites Based on PCL and Mesoporous Silicas: Effect of Pore Sizes and Their Influence in Ultimate Mechanical Response. *J. Compos. Sci.* **2021**, *5*, 321. [CrossRef]
32. Lommerse, J.P.M.; Price, S.L.; Taylor, R. Hydrogen bonding of carbonyl, ether, and ester oxygen atoms with alkanol hydroxyl groups. *J. Comput. Chem.* **1997**, *18*, 757–774. [CrossRef]
33. Pan, P.; Kai, W.; Zhu, B.; Dong, T.; Inoue, Y. Polymorphous Crystallization and Multiple Melting Behavior of Poly(L-lactide): Molecular Weight Dependence. *Macromolecules* **2007**, *40*, 6898–6905. [CrossRef]
34. Kawai, T.; Rahman, N.; Matsuba, G.; Nishida, K.; Kanaya, T.; Nakano, M.; Okamoto, H.; Kawada, J.; Usuki, A.; Honma, N.; et al. Crystallization and melting behavior of poly (L-lactic acid). *Macromolecules* **2007**, *40*, 9463–9469. [CrossRef]
35. Mandelkern, L. *Crystallization of Polymers*; McGraw-Hill: New York, NY, USA, 1964.
36. Hoffman, J.D.; Davis, G.T.; Lauritzen, J.I. *Treatise of Solid State Chemistry*; Plenum: New York, NY, USA, 1976.
37. Dobreva, T.; Pereña, J.M.; Pérez, E.; Benavente, R.; García, M. Crystallization behavior of poly(L-lactic acid) based eco-composites prepared with kenaf fiber and rice straw. *Polym. Composite* **2010**, *31*, 974–984. [CrossRef]

 *polymers*

Article

Hybrid Xerogels: Study of the Sol-Gel Process and Local Structure by Vibrational Spectroscopy

Guillermo Cruz-Quesada [1,2], Maialen Espinal-Viguri [1,2,*], María Victoria López-Ramón [3] and Julián J. Garrido [1,2,*]

1   Departamento de Ciencias, Campus Arrosadía, Public University of Navarre (UPNA), 31006 Pamplona, Spain; guillermo.cruz@unavarra.es
2   Institute for Advanced Materials and Mathematics (INAMAT²), Campus Arrosadía, Public University of Navarre (UPNA), 31006 Pamplona, Spain
3   Departamento de Química Inorgánica y Orgánica, Facultad de Ciencias Experimentales, University of Jaén, 23071 Jaén, Spain; mvlro@ujaen.es
*   Correspondence: maialen.espinal@unavarra.es (M.E.-V.); j.garrido@unavarra.es (J.J.G.); Tel.: +34-948-169604 (M.E.-V.); +34-948-169601 (J.J.G.)

Abstract: The properties of hybrid silica xerogels obtained by the sol-gel method are highly dependent on the precursor and the synthesis conditions. This study examines the influence of organic substituents of the precursor on the sol-gel process and determines the structure of the final materials in xerogels containing tetraethyl orthosilicate (TEOS) and alkyltriethoxysilane or chloroalkyltriethoxysilane at different molar percentages (RTEOS and ClRTEOS, R = methyl [M], ethyl [E], or propyl [P]). The intermolecular forces exerted by the organic moiety and the chlorine atom of the precursors were elucidated by comparing the sol-gel process between alkyl and chloroalkyl series. The microstructure of the resulting xerogels was explored in a structural theoretical study using Fourier transformed infrared spectroscopy and deconvolution methods, revealing the distribution of $(SiO)_4$ and $(SiO)_6$ rings in the silicon matrix of the hybrid xerogels. The results demonstrate that the alkyl chain and the chlorine atom of the precursor in these materials determines their inductive and steric effects on the sol-gel process and, therefore, their gelation times. Furthermore, the distribution of $(SiO)_4$ and $(SiO)_6$ rings was found to be consistent with the data from the X-ray diffraction spectra, which confirm that the local periodicity associated with four-fold rings increases with higher percentage of precursor. Both the sol-gel process and the ordered domains formed determine the final structure of these hybrid materials and, therefore, their properties and potential applications.

Keywords: ORMOSILs; xerogels; hybrid materials; chloroalkyltriethoxysilanes; inductive effect; $(SiO)_x$ structures; FTIR

Citation: Cruz-Quesada, G.; Espinal-Viguri, M.; López-Ramón, M.V.; Garrido, J.J. Hybrid Xerogels: Study of the Sol-Gel Process and Local Structure by Vibrational Spectroscopy. *Polymers* **2021**, *13*, 2082. https://doi.org/10.3390/polym13132082

Academic Editor: Jesús-María García-Martínez

Received: 31 May 2021
Accepted: 22 June 2021
Published: 24 June 2021

Publisher's Note: MDPI stays neutral with regard to jurisdictional claims in published maps and institutional affiliations.

1. Introduction

Organically modified silicon xerogels (ORMOSILs) are attracting considerable interest for their application in new-generation technologies, being utilized in chemical and optical sensors [1–6], for catalysis [4,5], in coatings [6–10], for chromatography [11,12], and in pharmacy [13]. They have great chemical versatility and can be efficiently modified for specific applications due to their combination of highly varied mechanical, optical and textural properties [14].

The sol-gel method is the most widely used approach to the synthesis of hybrid silicon xerogels. It is based on co-condensation between monomers of traditional alkoxides $(Si(OR)_4)$ such as tetramethoxy- or tetraethoxysilane (TEOS and TMOS, respectively) and one or more mono-, di- or trialkylsilanes $(R_xSi(OR')_{4-x})$ [15]. In hybrid xerogels, the addition of organic molecules or functional groups in the silica network restricts the three-dimensional growth of the material and blocks a condensation position, favoring the preferential formation of $(SiO)_4$ and $(SiO)_6$ rings in the amorphous structure of the silica materials [16,17] and even leading to the formation of structures with a higher degree of order [18]. However, although the sol-gel process has been widely studied [19], some questions have yet to be adequately settled. For instance, multiple reactions take place simultaneously in the process, making it difficult to extract information from experimental procedures [20]. To date, many studies have attempted to elucidate the influence of precursors on the hydrolysis, condensation reactions, and crosslinking by applying different characterization and analysis techniques, including gas chromatography (GC) [21], nuclear magnetic resonance ^{29}Si NMR [22,23], Raman spectroscopy [24], and Fourier transform infrared spectroscopy (FTIR) [25,26]. Among these approaches, FTIR not only provides complementary information on the bonds and structures formed during the synthesis process but also, when combined with deconvolution methods, yields valuable data on the microstructure of siliceous materials [27–29]. Knowledge obtained by these means is of major interest because it allows for the prediction of: (i) the physical properties of xerogels derived from their structure [30]; (ii) the catalytic activity of metals supported in silica matrices [31]; and (iii) the correlation of the silica species in the membrane of a fiber optic sensor with its efficacy [32].

The ultimate application of these materials is as membranes in fiber optic sensors, on which the analyte is specifically and reversibly adsorbed. Physisorption of the analyte on the surface of the material generates a modification of the refractive index and produces a change in the reflected optical power, which determines the analyte concentration in the medium. For this reason, it is important to prepare materials with different porosities and surface chemistries in which the interaction between the membrane (chemical area of a sensor) and the analyte is specific and labile. To date, our research group has prepared membranes with hybrid xerogels obtained using different molar ratios of TEOS:RTEOS (R = methyl or phenyl) [33,34]. Given the results obtained and following this line of reasoning, the addition of a chlorine atom to the organic moiety of a silane emerged as a highly appealing approach because of the inductive effects of the chlorine, which generates a more active chemical surface and favors functionalization with other compounds. For these reasons, ClRTEOS precursors were used to synthesize three series of xerogels analogous to those prepared in previous studies [35].

The objective of this study was to determine the influence of the alkyl and chloroalkyl substituents of the precursors on: (i) the gelation time, and (ii) the microstructure of the synthesized materials, obtained by deconvolution of the FTIR spectra. The study used six series of hybrid xerogels prepared (in previous studies) with tetraethoxysilane (TEOS) and a chloroalkyl or alkyl precursor at different molar percentages (ClRTEOS or RTEOS: R = methyl [M], ethyl [E]; and propyl [P]) [35–38]. Theoretical study of the deconvolution of the band at 1300–980 cm^{-1} in the FTIR spectra yielded semi-quantitative information on the proportion of $(SiO)_4$ and $(SiO)_6$ rings related to periodic structural domains and amorphous silica, respectively. This allows the local internal order of materials to be determined and the influences of the organic substituent and chlorine atom to be predicted. The ultimate application of these materials is as membranes in fiber optic sensors, constituting the chemical area of the sensor, where the analyte is specifically and reversibly adsorbed. Physisorption of the analyte on the surface of the material generates a modification of the refractive index and produces a change in the reflected optical power, which determines the analyte concentration in the medium. This study is of crucial importance because: (i) precise knowledge of the gelation time is essential to effectively impregnate the fibers [33,34], and (ii) complete characterization of the xerogels allows prediction, to a large extent, of their properties and the a priori selection of the xerogel best suited to the characteristics of the analyte of interest [32].

2. Materials and Methods

2.1. Materials

The siliceous precursors TEOS (tetraethoxysilane, purity > 99%), ClMTEOS (chloromethyl)triethoxysilane, purity > 95%), and ClPTEOS ((3-chloropropyl)triethoxy silane, purity > 95%) were supplied by Sigma-Aldrich (San Luis, MO, USA), and ClETEOS ((2-Chloroethyl)triethoxysilane, purity > 95%) was obtained from Fluorochem (Glossop, UK). Absolute ethanol (Emsure®) and hydrochloric acid (HCl, 37% *w/w*) were purchased from Merck (Darmstad, Germany) and potassium bromide (FT-IR grade) from Sigma-Aldrich (San Luis, MO, USA). All chemicals were used without further purification.

2.2. Synthesis of the Xerogels

The three series of hybrid xerogels were prepared as previously described [35–38]. TEOS was mixed with ClRTEOS (R = Methyl [M], Ethyl [E] or propyl [P]) at different molar ratios, maintaining a constant (TEOS + RTEOS):ethanol:water ratio (1:4.75:5.5) throughout the series.

The reagent and solvent quantities were adjusted to obtain 20 mL of alcogels. First, TEOS and ClRTEOS were mixed in a 30 mL container (φ 3.5 cm, threaded plastic lid, Schrarlab, Barcelona, Spain). Absolute ethanol was then added, followed by the dropwise addition of Milli-Q grade water under magnetic stirring to facilitate miscibility. Once the pH of the mixtures remained unchanged (after ~10 min), an automatic burette (Tritino mod. 702 SM, Metrohm, Herisau, Switzerland) was used to set the pH at 4.5 (0.05 M solution of HCl), and the mixture was stirred for 10 min to ensure homogenization. The closed containers were placed in a thermostatized oven at 60 °C (J.P. Selecta S.A, Barcelona, Spain) until gelling, i.e., when the shape of the materials did not change when the container was tilted. Subsequently, 5 mL of ethanol were added to cure the alcogel at room temperature for one week. Next, the containers were opened and covered with paraflim™, which was pierced with holes to facilitate evaporation of the solvent, and were then dried at room temperature under atmospheric pressure. The monolith was considered dried when no significant variation in its mass was observed. Finally, the xerogels were further dried (90 °C under vacuum) and then ground in an agate mortar.

2.3. Characterization

Infrared spectra were recorded using a FTIR spectrometer (Jasco, mod. 4700, Tokyo, Japan) at 25 scans and resolution of 4 cm^{-1}. Compressed KBr tablets with two different concentrations of sample were prepared: (i) 0.6 mg sample in 200 mg KBr tablet for spectra in the range 2200–400 cm^{-1}, thereby avoiding saturation of the Si-O-Si asymmetric stretching signal [39]; and (ii) 2 mg sample in 200 KBr tablet for spectra in the range 4000–2200 cm^{-1} for recording O-H and C-H bonds in greater detail. After their preparation, the tablets were dried over-night in an oven at 115 °C under vacuum to remove adsorbed water. Comparison between functional groups of the precursors in the pure reactant versus xerogel was carried out by using attenuated-total reflectance (ATR) to record the spectra directly from droplets of the precursors. The recorded transmittance of the samples was transformed into Kubelka-Munk units with the spectrometer software (Spectramanager, SMII FTIR Rev 216A ver2.15A) to allow their deconvolution (curve fitting) by a previously reported method [27]. Study parameters were four or five Gaussian-Lorentzian bands, with a maximum of 200 interactions, fixed baseline and >0.1 difference between experimental and synthetic spectra.

X-ray diffraction spectra were obtained at room temperature using a PANanalytical Empydrean XRD instrument (Empyrean, Almelo, The Netherlands) with graphite monochromator (at 45 KV and 40 mA) and copper rotating anode to select the CuK$_{\alpha 1/2}$ wavelength at 1.54 nm. Measurements were performed in a stepped scan mode of $2 \leq 2\theta \leq 50°$ in steps of 0.013° at a rate of 0.5 steps s^{-1} [40].

3. Results and Discussion

3.1. Influence of Organochlorine Substituents on Gelation Time

The gelation time is the period between the initial mixture of reagents and the formation of the gel; it comprises hydrolysis, condensation, and gelation (stabilization of colloids and crosslinking) stages, with gelation being the limiting stage [41]. Density functional theory (DFT) studies of TEOS in acidic media showed that the overall process of hydrolysis presents a pseudo-first order mechanism (SN_1) with lesser energy barriers in comparison to condensation [20]. Moreover, each consecutive hydrolysis reaction presents a barrier with less energy, which is consistent with the hydrolysis rates observed in ^{29}Si NMR experiments on various organoalkylsilanes [23]. The hydrolysis reactions of RTEOS are described in the following equations:

$$RSi(OC_2H_5)_3 + H_2O \xrightarrow{k_{h1}, \ [H_3O^+]} RSi(OC_2H_5)_2(OH) + C_2H_5OH, \tag{1}$$

$$RSi(OC_2H_5)_2(OH) + H_2O \underset{k_{-h2}, \ [H_3O^+]}{\overset{k_{h2}, \ [H_3O^+]}{\rightleftharpoons}} RSi(OC_2H_5)(OH)_2 + C_2H_5OH, \tag{2}$$

$$RSi(OC_2H_5)(OH)_2 + H_2O \underset{k_{-h3}, \ [H_3O^+]}{\overset{k_{h3}, \ [H_3O^+]}{\rightleftharpoons}} RSi(OH)_3 + C_2H_5OH. \tag{3}$$

Reaction rates in the acidic hydrolysis of triethoxysilanes depend on the competition between the steric hindrance and inductive effects of the organic moieties [16]. The reaction rate decreases with an increase in the volume of the organic moiety [42], while electron donating groups ($^+$I) stabilize the pentacoordinated transition state and increase the electronic charge of the ethoxide in the hybrid precursor, facilitating the attack of the acidic proton (reactions 1–3) [20,43]. In previous studies, it was confirmed that the hydrolysis rates increase with a longer alkyl chain in the precursor due to the enhanced inductive effect of the moieties [23]. Given that chloroalkyltriethoxysilanes contain chlorine, an electron withdrawing group ($^-$I), hydrolysis rates are expected to be lower than their non-chlorinated analogs, as can be deduced from the partial charges of the ethoxide groups (δ_{OEt} values of Table 1).

Table 1. Electronegativities and partial charges of the chloroalkyl precursors and their alkyl analogs calculated from the Pauli electronegativities and application of the equations of Livage and Henry [44].

Precursor	R	X	δ_{Si}	δ_R	δ_{OEt}
TEOS	OC_2H_5	2.32	0.32	-	−0.08
MTEOS	CH_3	2.29	0.31	0.20	−0.17
ETEOS	C_2H_5	2.29	0.31	0.28	−0.20
PTEOS	C_3H_7	2.28	0.30	0.35	−0.22
ClMTEOS	CH_2Cl	2.32	0.32	−0.09	−0.08
ClETEOS	C_2H_4Cl	2.31	0.32	0.02	−0.11
ClPTEOS	C_3H_6Cl	2.30	0.31	0.11	−0.14

X, molecule electronegativity; δ, partial charge.

However, the condensation reactions determine the overall speed of the gelation process. The following equations exhibit the SN_2 mechanism for the first and consecutive acidic condensation reactions of RTEOS:

$$RSi(OH)_3 + RSi(OH)_3 \xrightarrow{k_{c1}, \ [H_3O^+]} OH-RSi(OH)-O-RSi(OH)-OH + H_2O, \tag{4}$$

$$RSi(OH)_3 + (OH)_m(RSiORSi)_n \xrightarrow{k_{c2}, \ [H_3O^+]} (OH)_{m-1}(RSiORSi)_{n+1} + H_2O. \tag{5}$$

In contrast to the hydrolysis reactions (Reactions 1–3), the first condensation between hydrolyzed molecules is kinetically favored, leading initially to the formation of long and slightly branched chains and subsequently to intra-molecular condensations and cyclization [16,45,46]. The successive condensations form the polymeric network and the first colloidal particles (φ = 10–100 nm), which give rise to a spontaneous and homogeneous nucleation process when their critical radius is reached. It should be noted that gelation of the hybrid material is slower because triethoxysilanes have only three potential directions of polymerization. Unlike in the hydrolysis process, where the more the substituent of the precursor is withdrawn, the more favored is the condensation, because it favors nucleophilic attack on the silicon atom of a neighboring molecule. In this way, the condensation rate increases in the order MTEOS > ETEOS > PTEOS for the alkyl series and ClMTEOS > ClETEOS > ClPTEOS for the chloroalkyl series. In addition, chloroalkyl groups restrict the crosslinking between oligomers due to electrostatic repulsions, which destabilize the colloids and prevent the gelling process above a given molar percentage. Thus, it was possible to synthesize hybrid xerogels with molar percentages of up to 35%, 25% and 25% for the chloroalkyl series (ClMTEOS, ClETEOS and ClPTEOS respectively, Figure S1), while the molar percentages were substantially higher for the alkyl series [36–38].

Hence, with regard to the gelation times of the synthesized materials, there is a balance between the inductive/steric effects of the organic substituents of the precursor and the electrostatic effect between colloids. This electrostatic effect was a determining factor in the gelation times of chloroalkyl xerogels with a high percentage of precursor, due to the chlorine atom. The corresponding data are displayed in Figure 1a, which depicts the variation in gelation times of the chloroalkyl xerogels as a function of the molar percentage of precursor. It should be noted that gelation times for percentages higher than 10% have been omitted for the ClPTEOS series because they do not follow the trend observed for 0 to 10 molar percentages due to the limitations imposed by steric and electrostatic factors. In all cases, an increase in the molar percentage lengthens the gelation time, and this trend is also observed in xerogels prepared with analogous alkyl precursors, as shown in Figure 1b [36–38]. Figure 1c compares gelation times between the chloroalkyl xerogels and their non-chlorinated analogs, showing that the effect of the chlorine atom is more pronounced in the ClETEOS and ClPTEOS series, besides being opposite to the effect in the ClMTEOS series. This is because the alkyl chain is longer in the ClETEOS and ClPTEOS, increasing the steric effect and the repulsions between colloids created by the chlorine atom, which markedly lengthens the gelation time in comparison to their analogous non-chlorinated series. The chain is shorter in the ClMTEOS series; therefore, the steric effect is less pronounced and the repulsion between colloids is lower, favoring crosslinking in comparison to the longer chain chloroalkyl series. Furthermore, the withdrawing effect of the chlorine atom on the adjacent silicon atom is maximized, favoring condensation. Accordingly, the gelation times for the ClMTEOS series are reduced with respect to the analogous non-chlorinated series.

It should be noted that t_g values are higher for the ClETEOS series than for the other two chloroalkyl series and fit a linear rather than exponential trend (Figure 1a). This is because the length of the chloroethyl chain is not as short as in the ClMTEOS series or as long as in the ClPTEOS series, which has greater flexibility and freedom of movement. As demonstrated below, the behavior of this substituent is different because its size and nature necessarily place it in the network at fixed positions, minimizing repulsions. The lack of freedom of movement hinders and slows crosslinking, markedly increasing gelation times and forming periodic structures that create ordered domains in the silica network, as explained below.

3.2. Study of the Local Structure of Hybrid Xerogels Using FTIR and XRD

Figure 2 depicts, as an example, the FTIR spectra of the three chloroalkyl series with 20% molar percentage of the precursor in two wavelength ranges: (i) 4000–2750 cm^{-1} and (ii) 1600–400 cm^{-1}. The spectra obtained for the xerogels prepared with different molar percentages of chloroalkyl precursor and the different modes of vibration are reported in Supplementary Material with corresponding citations from the literature (Figure S2, Table S1).

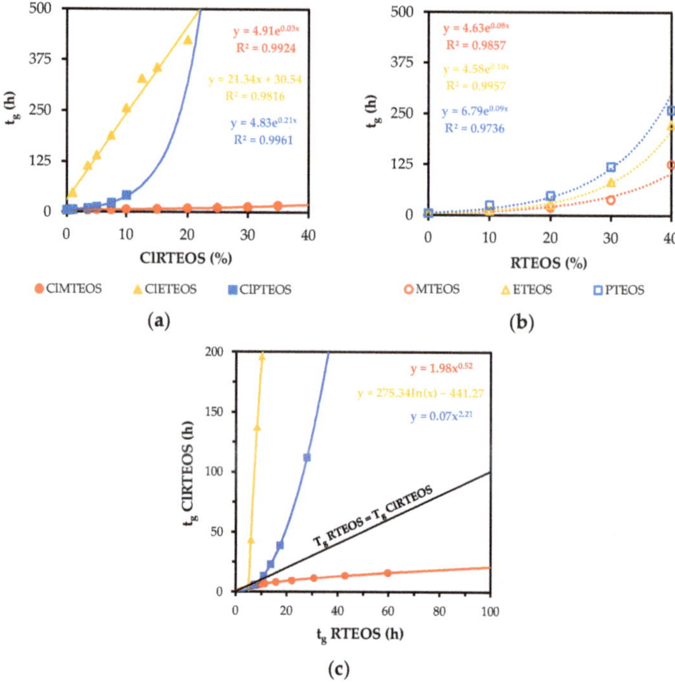

Figure 1. (**a**) t_g as a function of the percentage of precursor in the TEOS:ClRTEOS series, (**b**) t_g as a function of the percentage of t precursor in the TEOS:RTEOS series, and (**c**) t_g of the ClRTEOS versus RTEOS series.

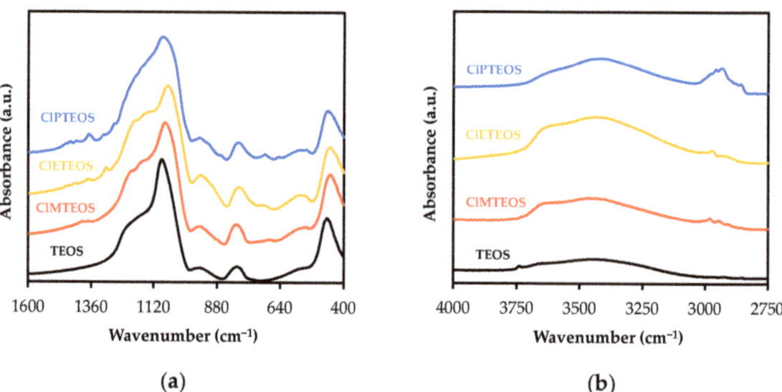

Figure 2. FTIR spectra of TEOS xerogel and TEOS:ClRTEOS hybrids (molar ratio, 20%): (**a**) range from 1600 to 400 cm^{-1}, and (**b**) range from 4000 to 2750 cm^{-1}.

Figure 2a depicts the characteristic FTIR bands of the silica lattice: (i) rocking of O-Si-O (ρ O-Si-O) at 455 cm^{-1}, (ii) symmetric stretching vibration of Si-O-Si (ν_s Si-O-Si) at 800 cm^{-1}, (iii) stretching of the Si-O bond belonging to surface silanes (ν_s Si-OH) at 955 cm^{-1}, (iv) asymmetric stretching vibration Si-O-Si at 1090 cm^{-1} (ν_{as} Si-O-Si), and (v) a broad and intense shoulder at 1200 cm^{-1} related to various modes of vibration of the Si-O-Si bonds [47]. Additionally, materials synthesized using silsesquioxanes typically show a set of Si-O-Si vibrations belonging to different structures: (i) bicyclic species at 1007 cm^{-1}, (ii) linear species at 1020 cm^{-1}, and (iii) cyclic species at 1050 cm^{-1} [48]. As can be observed in the figure, bands associated with bicyclic and linear species do not appear to overlap with the band at 1090 cm^{-1}, indicating that these structures are present in lower proportions than are the cyclic structures. Figure 2b displays the bands of isolated (or non-interacting) surface silanols and silanols interacting by hydrogen bonds at 3660 cm^{-1} and 3450 cm^{-1}, respectively. These bands, which are also characteristic of silicon xerogels, indicate the hydrophobic or hydrophilic nature of the material [49]. In addition to the aforementioned bands, a shoulder is observed at 550 cm^{-1} in all spectra, associated with the presence of 4-membered siloxane rings, (SiO)$_4$ [50,51], which is consistent with the aforementioned assumption of the predominance of cyclic species.

The presence of organic groups in these hybrid materials is revealed by the bond vibrations of the alkyl chain: (i) C-H, (ii) C-C; and (iii) C-Cl. In Figure 2b, the characteristic bands of symmetric and asymmetric stretching of C-H are clearly observed between 2890 and 2975 cm^{-1}, showing an increase in intensity with longer chain length from being practically null in TEOS to being readily identifiable in ClPTEOS. The same can be seen in Figure 2a for the region between 1250 and 1400 cm^{-1}, which shows the bands of deformation modes related to C-H bonds (δ C-H) [52]. The symmetric stretching band C-C (ν_s C-C) of chloroethyl was detected in ClETEOS series at 900 cm^{-1} [53], while two well-differentiated bands were detected at 920 and 870 cm^{-1} in the ClPTEOS series, corresponding to the C-C bond contiguous to the chlorine and silicon atom, respectively (Table S1). This figure also depicts C-Cl stretching vibration bands (ν C-Cl) in the region between 750 and 650 cm^{-1} [53,54], with an increased intensity at higher molar percentages of ClRTEOS, as expected (Figure S2). It should be noted that it was not possible to detect the spectral band of the wagging vibration of the Si-CH$_2$ bond (ω C-H) in the region between 1300 and 980 cm^{-1} and the band corresponding to the out-of-plane deformation of the Si-C bond due to overlap with the frequencies of the bending vibration of the Si-O bond at 800 cm^{-1}. However, both vibrations can be clearly observed in the attenuated total reflection (ATR) spectrum of the pure precursor (Figure S3).

In the alkyl series previously studied by our research group (RTEOS:TEOS, R = M, E or P), a splitting of the band is observed at high percentages of alkyl precursor, moving the original band to lower frequency values due to the inductive effect of alkyl groups [36–38]. In the ETEOS series, for example, the band at 1092 cm^{-1} shifts to 1043 cm^{-1} and a new band appears at 1128 cm^{-1}, increasing its relative absorbance with higher molar percentage of ETEOS. The appearance of this band is related to the presence of highly symmetric structures within the amorphous silica matrix (Figure 3). These structures comprise four-membered silicon rings (SiO)$_4$ and are described as close or open cages (T$_8$ and T$_7$, respectively) and short ladders [54,55]. The intensity of this new band increases with higher molar percentage of the chloroalkyl precursor due to stabilization of the rings (SiO)$_4$ and minimization of the electrostatic repulsions produced by the chlorine atoms. In the chloroalkyl series, the asymmetric Si-O-Si vibration also shifts towards lower frequency values with increased proportion of the precursor. The bands are not resolved in this case because the molar percentage reached is lower than in the alkyl series; instead, shoulders are observed at around 1200 cm^{-1}.

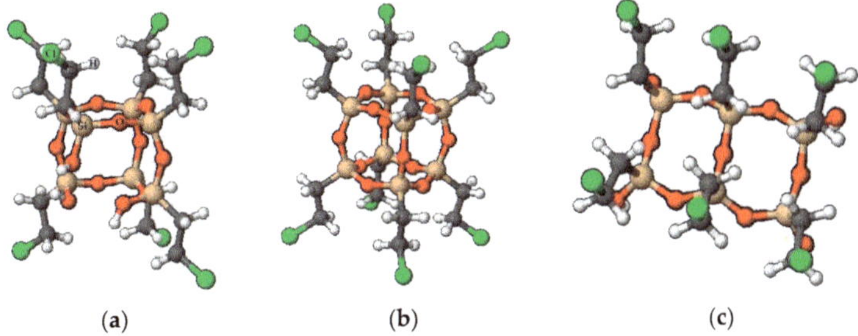

(a)   (b)   (c)

Figure 3. Ordered structures in the silica matrix built by (SiO)$_4$ rings in ClETEOS as an example: (a) open cage (T$_7$), (b) cage (T$_8$), and (c) short ladders.

The presence of these structures in both RTEOS:TEOS and ClRTEOS:TEOS hybrid xerogels is consistent with the X-ray diffraction spectra obtained for these materials. Figure 4 depicts the X-ray diffraction spectra of the ClETEOS:TEOS series (the only series that follows a linear trend in gelation times). In addition to the characteristic diffraction peak of amorphous silica ($2\theta \sim 24$) [56,57], another peak can be observed at $2\theta < 10°$ with only 1% of precursor. This peak is related to cage-like and ladder-like oligomeric species that form ordered domains within the amorphous matrix of the xerogel [58,59], which, according to computational chemistry studies, are the most thermodynamically stable structures for MTEOS:TEOS hybrid xerogels at pH 4.5 [60]. In the analogous ETEOS series, this diffraction peak is detected above 30% ETEOS and is consistent with the mass spectrometry results, which indicate the presence of ladder-like structures within the silica matrix [37]. Consequently, the appearance of this peak at much lower molar percentages in the ClETEOS series can be explained by the steric and electronic properties of the chloroethyl substituent, conferring rigidity or freedom of movement restriction to the organic moiety. This compromises the crosslinking and encourages the early formation of ordered periodic domains in the 3D structure. The difficult cross-linking of the colloids or oligomers would explain the anomalous trends in gelation times for this series. The spectra of the series of hybrid xerogels and the corresponding data are exhibited in Supplementary Material (Figure S4, Table S2).

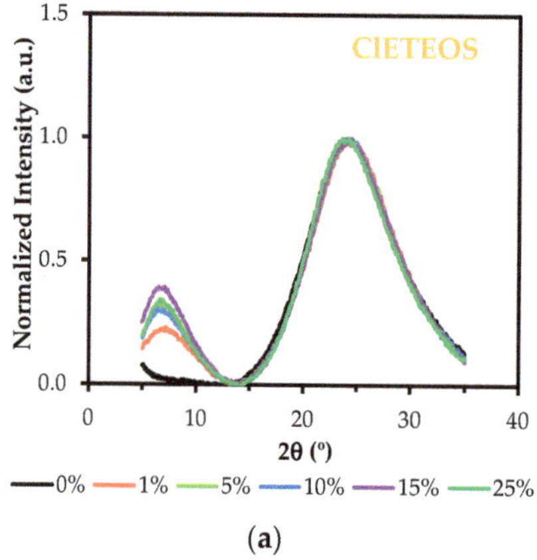

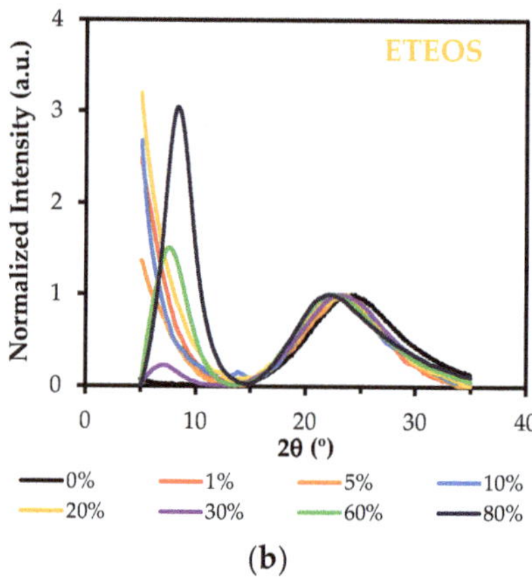

Figure 4. X-ray diffraction spectra of the hybrid xerogels at different molar percentages (normalized with respect to the band $2\theta \sim 4°$: (**a**) ClETEOS:TEOS, and (**b**) ETEOS:TEOS.

3.3. Spectral Deconvolution of the 1300–700 cm^{-1} Region

Figure 5 depicts the ν_{as} Si-O-Si band at 1090 cm^{-1} and the associated shoulder at around 1200 cm^{-1}. It shows the shift at lower frequencies of the band at 1090 cm^{-1} with higher molar percentages of the precursor as well as the overlapping bands derived from the shoulder at 1200 cm^{-1}. These emerging bands are attributable to the different structures that make up the siloxane bonds, mostly forming rings of four or six silicon atoms, $(SiO)_4$ and $(SiO)_6$ respectively. The relative abundance of these rings depends on the nature and molar ratio of the precursor: four-fold rings are the predominant species in xerogels and are thermodynamically favored in the oligomerization of TEOS at acid pH through cyclodimerization processes [45,47], whereas six-fold rings are kinetically favored over four-membered rings and constitute the main structures of silicates and amorphous silica [17,61,62]. In the deconvolution study, each structure is associated with two optical modes of ν_{as} Si-O-Si vibration in the FTIR spectra due to the Coulomb interactions: (i) transverse mode (TO), between 1100 and 1000 cm^{-1}; and (ii) longitudinal mode (LO), between 1250 and 1100 cm^{-1} [63].

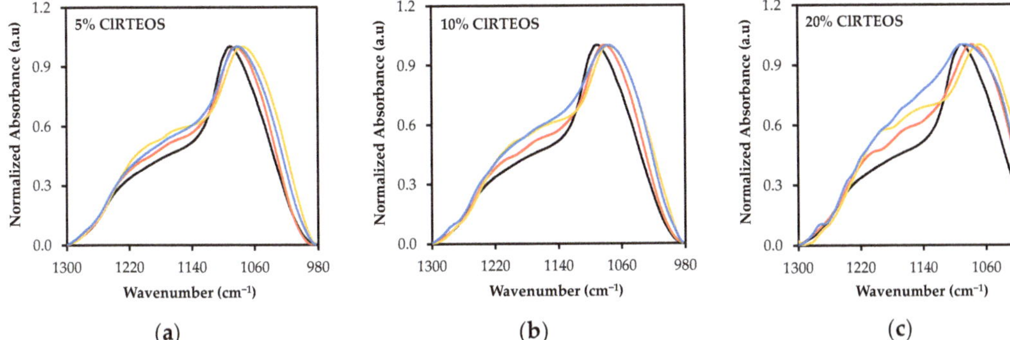

Figure 5. Normalized FTIR spectra of the 1300–980 cm^{-1} region of the TEOS xerogel (black), and those of the three chloroalkyl series at different molar ratios: (**a**) 5%, (**b**) 10%, and (**c**) 20%. (ClMTEOS (red), ClETEOS (yellow), and ClPTEOS (blue)).

The changes in these structures as the precursor increases and their relative proportions were studied by deconvolution of the FTIR spectra in the range 1300–980 cm^{-1}, using the non-linear least-squares method to obtain the Gaussian-Lorentzian components. The different distances and degrees of torsion of Si-O-Si bonds in the structures, along with the optical modes, predict four components of ν_{as} Si-O-Si in the studied range: (i) TO$_4$ and LO$_4$ for (SiO)$_4$ rings, and (ii) TO$_6$ and LO$_6$ for (SiO)$_6$ rings. Therefore, four components were introduced in the software for deconvolution of the reference material spectrum (100% TEOS Xerogel, Figure 2). The resulting optimized synthetic spectrum was composed of bells with maxima at 1214, 1143, 1092 and 1078 cm^{-1}, assigned to LO6, LO$_4$, TO$_4$ and TO$_6$, respectively. This assignation takes account of: (i) the aforementioned frequency ranges of the optical modes; (ii) the predominance of (SiO)$_4$ species evidenced by the shoulder at 550 cm^{-1} in the spectra, and (iii) the broader limiting frequencies in both optical modes for six-fold rings, related to their less tensioned nature (Figure 6) [27,47,64].

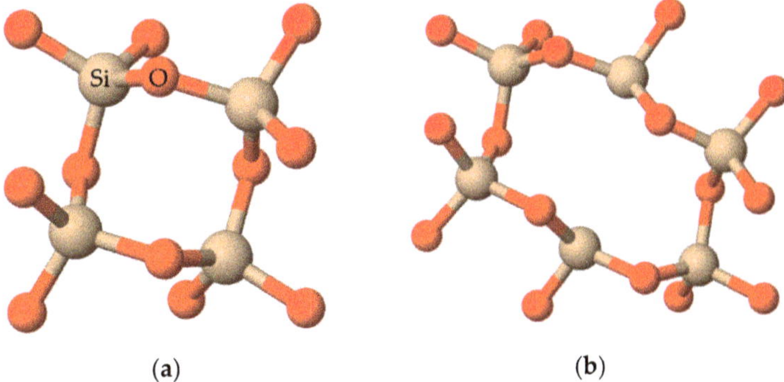

(a) (b)

Figure 6. Predominant structures within the silica matrix of a xerogel prepared with TEOS: (**a**) (SiO)$_4$ rings, and (**b**) (SiO)$_6$ rings.

Spectra of the RTEOS:TEOS and ClRTEOS:TEOS series were deconvoluted using the aforementioned frequencies obtained from the reference material (LO$_6$, LO$_4$, TO$_4$ and TO$_6$) and the frequencies of C-H wagging vibrations corresponding to the organic moiety of each precursor (Table S1, supporting material). By way of example, Figure 7 depicts the synthetic spectra derived from the bell-shaped curves obtained for TEOS (Figure 7a) and the organochlorinated xerogels with a 20% molar percentage of precursor (actual FTIR spectra shown in Figure 5c).

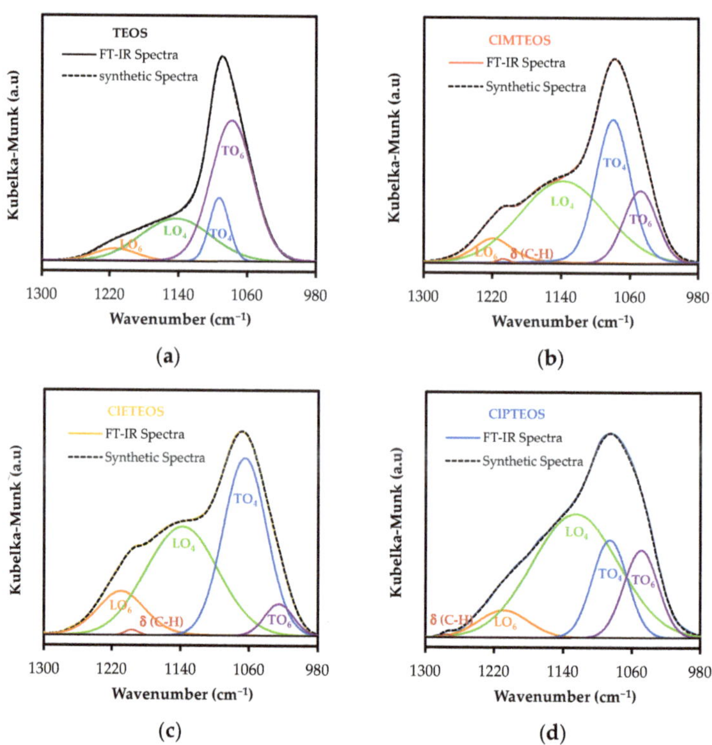

Figure 7. Deconvolution and least-squares adjustment of the FTIR spectra of the xerogels: (**a**) Reference material (100% TEOS), (**b**) 20% ClMTEOS, (**c**) 20% ClETEOS, and (**d**) 20% ClPTEOS.

Based on the integrated areas of the Gaussian-Lorentzian bells, Equations (6) and (7) were calculated to determine the relative abundance of $(SiO)_6$ and $(SiO)_4$ rings (Tables S3 and S4 for the ClRTEOS:TEOS and RTEOS:TEOS series, respectively).

$$(SiO)_6, \% = \frac{A(LO_6) + A(TO_6)}{A(LO_4) + A(TO_4) + A(LO_6) + A(TO_6)} \, 100, \tag{6}$$

$$(SiO)_4, \% = \frac{A(LO_4) + A(TO_4)}{A(LO_4) + A(TO_4) + A(LO_6) + A(TO_6)} \, 100, \tag{7}$$

where $A(LO_6)$ is the area of the band at 1220 cm^{-1}, $A(LO_4)$ is the area of the band at 1150 cm^{-1}; $A(TO_4)$ is the area of the band at 1070 cm^{-1} and $A(TO_6)$ the area of the band at 1050 cm^{-1}.

Figure 8 graphically depicts the $(SiO)_4/(SiO)_6$ ratio as a function of the molar percentage of chloroalkyl precursor (data obtained from Tables S3 and S4, see Supplementary Material). It shows that the formation of $(SiO)_4$ rings is more favored with longer chloroalkyl chain for a given molar percentage of precursor. The $(SiO)_4/(SiO)_6$ ratio increases exponentially with a higher percentage of precursor in the three series. These trends are not expected, because the entry of organic molecules or substituents into the network should produce an increase in the proportion of $(SiO)_6$ species, which minimizes steric tension, increases the volume, and markedly reduces the density (Table S5). Accordingly, Fidalgo, A. et al. found that the proportion of $(SiO)_6$ rings increases with higher molar percentage of RTEOS (R = M, E and P), reaching an abundance of almost 85% for samples with a precursor percentage of 75% [27]. However, the synthesis of these materials was performed in basic medium, explaining the greater abundance of six-membered rings from crosslinking between more branched chains. In the case of the present xerogels, the higher proportion of $(SiO)_4$ rings in the chloroalkyl series would be related not only to their synthesis in acidic medium but also to the formation of ordered domains at very low molar percentages of precursor. The data in Table S3 (Supplementary Material) for the three chloroalkyl series show that the contribution of the LO_4 band (attributable to the total percentage of $(SiO)_4$ rings) increases with higher percentage of precursor. This is a relevant finding, given that this band is also associated with the presence of ordered structures in the literature [5,51,65]. Its presence is closely related to a decrease in the degree of crosslinking in the xerogel, in agreement with the shifts at lower wavelengths of the TO_4 band, which is associated with the silicon network [13]. Another factor that supports the formation of these ordered structures is the decrease in the intensity of the LO_6 band (see Table S3), because this bands is generally associated not only with $(SiO)_6$ rings but also with the non-silica porous skeleton [66].

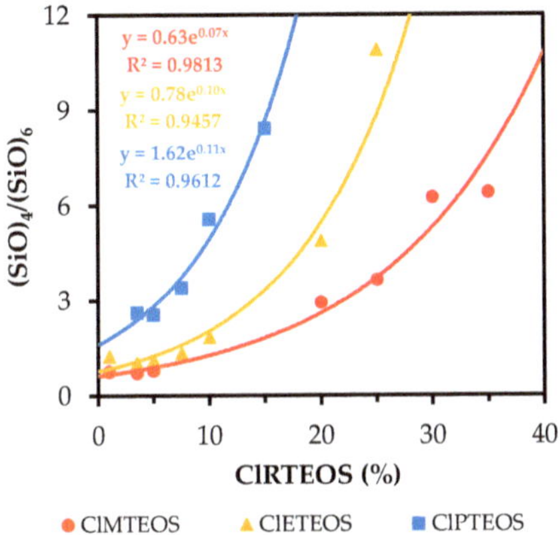

Figure 8. Variation in $(SiO)_4/(SiO)_6$ as a function of the molar percentage of precursor in ClR-TEOS:TEOS xerogels.

The way in which the chlorine atom affects the ring distribution was studied in greater depth by performing interpolations using the equations of the exponential curves shown in Figure 8 and Figure S5 ($(SiO)_4/(SiO)_6$ vs. %RTEOS). These equations were used to obtain $(SiO)_4/(SiO)_6$ ratios for percentages of the precursor in the range 1–35% in both the chlorinated series and their non-chlorinated analogs (the data used are exhibited in Table S6 of Supplementary Material).

Figure 9 compares $(SiO)_4/(SiO)_6$ ratios between the ClRTEOS:TEOS series and the RTEOS:TEOS series to show how the chlorine atom affects the relative proportion of species.

A discontinuous black straight line has been added in this figure to depict the hypothetical case in which the ratio of $(SiO)_4/(SiO)_6$ rings is the same for the alkyl and chloroalkyl series at the same molar percentage of precursor, which would indicate that the chlorine atom in the ClRTEOS precursor has no effect on the $(SiO)_4{:}(SiO)_6$ ratio with respect to its counterpart, RTEOS. Above this line, the effect of the chlorine on the $(SiO)_4{:}(SiO)_6$ ratio is positive, favoring the formation of $(SiO)_4$ rings. Below this line, the effect of the chlorine is negative, favoring the formation of $(SiO)_6$ structures. The figure shows that this ratio is lower in the ClMTEOS than in the MTEOS series. This trend is consistent with the behavior observed in Figure 1c, which shows that the ClMTEOS series has shorter gelation times, closely related to the size of the chloroalkyl chain. This not only fails to produce a significant steric or electrostatic effect to disfavor the formation of $(SiO)_6$ rings (kinetically favored species), but also accelerates condensation due to the inductive effect of the chlorine, which reduces gelation times and gives rise to lower $(SiO)_4/(SiO)_6$ ratios. Unlike the ClMTEOS series, the $(SiO)_4/(SiO)_6$ values for the ClETEOS and ClPTEOS series are higher than those of their analogs throughout the region, which is in turn consistent with the longer gelation times displayed in Figure 1c. In these two series, the steric factor and the electrostatic repulsions between colloids significantly increase the crosslinking time, disfavoring $(SiO)_6$ ring formation. The slope of these curves is much steeper than the slope observed for ClMTEOS series, implying that the addition of a higher percentage of precursor increases the influence exerted by the chlorine atom on the structure of the synthesized materials, which is maximized in ClPTEOS:TEOS, the series with the longest chain. It should be noted that, although the $(SiO)_4/(SiO)_6$ ratio is slightly lower in the ClETEOS versus ClPTEOS series, the former has longer gelation times and peaks at small angles ($10° < 2\theta$) in all X-ray diffraction spectra. According to these findings, the chloroethyl moiety in this material is more efficient in directing the formation of ordered domains and nanostructuring the material, even when it has fewer $(SiO)_4$ rings (related to periodic box or ladder-type structures), consistent with the results of all techniques used to characterize these hybrid materials [35].

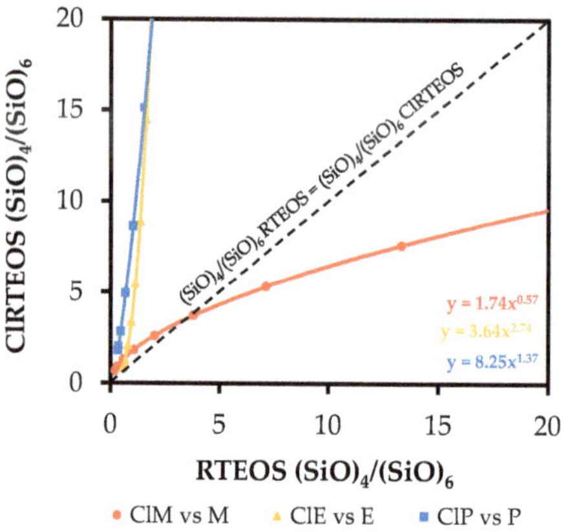

Figure 9. Variation in the proportion of structures $(SiO)_4/(SiO)_6$ of the chloroalkyl series with respect to the alkyl series at different molar percentages of precursor.

4. Conclusions

In general, an increase in the molar percentage of the precursor (ClRTEOS and RTEOS) translates into an increase in gelation times due to conjugation of the steric and inductive effects of the organic substituents. The gelation times of the chloroalkyl series are longer than those of their analogous alkyl series for long alkyl chains (ClETEOS and ClPTEOS), although they are slightly shorter for the shortest alkyl chain (ClMTEOS). This is explained by the maximization of the inductive effect of the chlorine atom due to its proximity to the silicon atom, which exerts a positive influence on the condensation and crosslinking reactions, the slowest step in the sol-gel process. On the other hand, a larger number of carbons in the alkyl chain reduces the inductive effect of the chlorine on the silicon atom and increases the steric effect exerted by the chain to form the silica network; this effect is maximized in the series bearing ethyl, chloroethyl, propyl and chloropropyl groups. Unexpectedly, the ClETEOS series have the highest gelation times, explained by: (i) the lack of flexibility of the chloroethyl group due to the steric tension caused by its size; (ii) the electrostatic repulsions exerted by chlorine, forcing the formation of kinetically disfavored structures in order to minimize its energy; and (iii) the X-ray diffraction spectra of this series (1–25%), which reveal maximum diffraction at a small angle ($2\theta < 10°$), consistent with the presence in their structure of periodic domains that require more time for their formation. The FTIR spectra of each hybrid xerogel yielded data for a semi-quantitative determination of the proportions of $(SiO)_4$ and $(SiO)_6$ rings by deconvolution methods. These findings reveal a competitive process between the two species that is dependent on the precursor and its molar ratio, observing similar trends to those obtained for the gelation times. In the ClMTEOS series, the presence of the chlorine atom favors the formation of six-fold rings. An opposite trend is observed in the ClETEOS and ClPTEOS series, favoring the formation of 4-fold rings; in addition, the chlorine atom exerts a stronger influence in the ClETEOS and ClPTEOS series, consistent with their longer gelation times and the presence of periodic structures. According to these results, the substitution of a hydrogen atom by a halogen functional group (e.g., chlorine) in hybrid xerogels produces relevant changes in its microstructure due to the intermolecular forces generated by the chlorine atom in the sol-gel process.

Supplementary Materials: The following are available online at https://www.mdpi.com/article/10.3390/polym13132082/s1, Figure S1: Images of the synthesized xerogels in which a change in morphology is observed with an increasing percentage of organic precursors: (**a**) TEOS, (**b–d**) ClM-TEOS, (**e–g**) ClETEOS, and (**h–j**) ClPTEOS; Figure S2: FTIR spectra of the ClRTEOS:TEOS xerogels in the ranges 4000–2750 and 1600–400 cm^{-1}: (**a,b**) ClMTEOS, (**c,d**) ClETEOS, and (**e,f**) ClPTEOS; Figure S3: FTIR spectra of the precursors (TEOS, ClMTEOS, ClETEOS and ClPTEOS) in the ranges: (**a**) 4000–2750 cm^{-1} and (**b**) 1600–400 cm^{-1}; Figure S4: X-ray diffraction spectra of ClRTEOS:TEOS and RTEOS:TEOS xerogels: (**a**) ClMTEOS, (**b**) MTEOS, (**c**) ClPTEOS, and (**d**) PTEOS; Figure S5: $(SiO)_4/(SiO)_6$ ratio with respect to the molar percentage of RTEOS:TEOS xerogels. Table S1: FTIR spectra band assignment; Table S2: Bragg angles (2θ), band area (A), and bond distance (d1 and d2 (nm)) calculated from the X-ray diffraction peaks for the xerogels synthesized with ClMTEOS, MTEOS, ClPTEOS, and PTEOS; Table S3: Relative areas obtained from deconvolution of the FTIR spectra and percentages of $(SiO)_4$ and $(SiO)_6$ in the ClRTEOS:TEOS series; Table S4: Relative areas obtained from deconvolution of the FTIR spectra and percentages of $(SiO)_4$ and $(SiO)_6$ in the RTEOS:TEOS series; Table S5: Helium density of the hybrid xerogels; Table S6: experimental data of $(SiO)_4/(SiO)_6$ and equations used for Figure 8 and Table 1.

Author Contributions: G.C.-Q.: investigation, writing–original draft. M.E.-V.: methodology, resources, writing, review and editing. M.V.L.-R. writing–review and editing. J.J.G.: conceptualization, supervision, project administration, funding acquisition, writing, review and editing. All authors have read and agreed to the published version of the manuscript.

Funding: This research was funded by the Ministerio de Economía y Competitividad (Project ref. MAT2016-78155-C2-2-R).

Data Availability Statement: The data presented in this study are available on request from the corresponding author.

Acknowledgments: The authors gratefully acknowledge the financial support received from the Ministerio de Economia y Competitividad of Spain (Project MAT2016-78155-C2-2-R). G.C. thanks MINECO and the "Ministerio de Ciencia, Investigación y Universidades" of Spain for his "FPU" grant (FPU18/03467). The authors also acknowledge the use of the "Centro de Instrumentación Científico-Técnica" at the University of Jaén and UCTAI at the Public University of Navarre.

Conflicts of Interest: The authors declare no conflict of interest.

References

1. Pastore, A.; Badocco, D.; Pastore, P. Influence of surfactant chain length, counterion and OrMoSil precursors on reversibility and working interval of pH colorimetric sensors. *Talanta* **2020**, *212*, 120739. [CrossRef]
2. Gillanders, R.N.; Campbell, I.A.; Glackin, J.M.E.; Samuel, I.D.W.; Turnbull, G.A. Ormosil-coated conjugated polymers for the detection of explosives in aqueous environments. *Talanta* **2018**, *179*, 426–429. [CrossRef]
3. Li, Z.; Suslick, K.S. Ultrasonic Preparation of Porous Silica-Dye Microspheres: Sensors for Quantification of Urinary Trimethylamine N-Oxide. *ACS Appl. Mater. Interfaces* **2018**, *10*, 15820–15828. [CrossRef]
4. Shamir, D.; Elias, I.; Albo, Y.; Meyerstein, D.; Burg, A. ORMOSIL-entrapped copper complex as electrocatalyst for the heterogeneous de-chlorination of alkyl halides. *Inorg. Chim. Acta* **2020**, *500*, 119225. [CrossRef]
5. Ponamoreva, O.N.; Afonina, E.L.; Kamanina, O.A.; Lavrova, D.G.; Arlyapov, V.A.; Alferov, V.A.; Boronin, A.M. Yeast Debaryomyces hansenii within ORMOSIL Shells as a Heterogeneous Biocatalyst. *Appl. Biochem. Microbiol.* **2018**, *54*, 736–742. [CrossRef]
6. Lin, W.; Zhang, X.; Cai, Q.; Yang, W.; Chen, H. Dehydrogenation-driven assembly of transparent and durable superhydrophobic ORMOSIL coatings on cellulose-based substrates. *Cellulose* **2020**, *27*, 7805–7821. [CrossRef]
7. Liu, Z.; Tian, S.; Li, Q.; Wang, J.; Pu, J.; Wang, G.; Zhao, W.; Feng, F.; Qin, J.; Ren, L. Integrated Dual-Functional ORMOSIL Coatings with AgNPs@rGO Nanocomposite for Corrosion Resistance and Antifouling Applications. *ACS Sustain. Chem. Eng.* **2020**, *8*, 6786–6797. [CrossRef]

8. Scotland, K.M.; Shetranjiwalla, S.; Vreugdenhil, A.J. Curable hybrid materials for corrosion protection of steel: Development and application of UV-cured 3-methacryloxypropyltrimethoxysilane-derived coating. *J. Coat. Technol. Res.* **2020**, *17*, 977–989. [CrossRef]

9. Bouvet-Marchand, A.; Graillot, A.; Abel, M.; Koudia, M.; Boutevin, G.; Loubat, C.; Grosso, D. Distribution of fluoroalkylsilanes in hydrophobic hybrid sol-gel coatings obtained by co-condensation. *J. Mater. Chem. A* **2018**, *6*, 24899–24910. [CrossRef]

10. Yue, D.; Feng, Q.; Huang, X.; Zhang, X.; Chen, H. In situ fabrication of a superhydrophobic ORMOSIL coating on wood by an ammonia-HMDS vapor treatment. *Coatings* **2019**, *9*, 556. [CrossRef]

11. Malek, S.K.; Nodeh, H.R.; Akbari-Adergani, B. Silica-based magnetic hybrid nanocomposite for the extraction and preconcentration of some organophosphorus pesticides before gas chromatography. *J. Sep. Sci.* **2018**, *41*, 2934–2941. [CrossRef] [PubMed]

12. Moriones, P.; Ríos, X.; Echeverría, J.C.; Garrido, J.J.; Pires, J.; Pinto, M. Hybrid organic-inorganic phenyl stationary phases for the gas separation of organic binary mixtures. *Colloids Surf. A Physicochem. Eng. Asp.* **2011**, *389*, 69–75. [CrossRef]

13. Tran, H.N.; Nghiem, T.H.L.; Vu, D.T.T.; Pham, M.T.; Nguyen, T.V.; Tran, T.T.; Chu, V.H.; Tong, K.T.; Tran, T.T.; Le, X.T.T.; et al. Dye-doped silica-based nanoparticles for bioapplications. *Adv. Nat. Sci. Nanosci. Nanotechnol.* **2013**, *4*, 043001. [CrossRef]

14. Judeinstein, P.; Sanchez, C. Hybrid organic-inorganic materials: A land of multidisciplinarity. *J. Mater. Chem.* **1996**, *6*, 511–525. [CrossRef]

15. Alemán, J.; Chadwick, A.V.; He, J.; Hess, M.; Horie, K.; Jones, R.G.; Kratochvíl, P.; Meisel, I.; Mita, I.; Moad, G.; et al. Definitions of terms relating to the structure and processing of sols, gels, networks, and inorganic-organic hybrid materials (IUPAC recommendations 2007). *Pure Appl. Chem.* **2007**, *79*, 1801–1829. [CrossRef]

16. Brinker, C.J.; Scherer, G.W. *Sol-Gel Science*; Academic Press: New York, NY, USA, 1990; ISBN 978-0-08-057103-4.

17. Fidalgo, A.; Ilharco, L.M. The defect structure of sol-gel-derived silica/polytetrahydrofuran hybrid films by FTIR. *J. Non Cryst. Solids* **2001**, *283*, 144–154. [CrossRef]

18. Chemtob, A.; Ni, L.; Croutxé-Barghorn, C.; Boury, B. Ordered hybrids from template-free organosilane self-assembly. *Chem. A Eur. J.* **2014**, *20*, 1790–1806. [CrossRef] [PubMed]

19. Issa, A.A.; Luyt, A.S. Kinetics of alkoxysilanes and organoalkoxysilanes polymerization: A review. *Polymers* **2019**, *11*, 537. [CrossRef] [PubMed]

20. Cheng, X.; Chen, D.; Liu, Y. Mechanisms of silicon alkoxide hydrolysis-oligomerization reactions: A DFT investigation. *ChemPhysChem* **2012**, *13*, 2392–2404. [CrossRef]

21. Issa, A.A.; El-Azazy, M.; Luyt, A.S. Kinetics of alkoxysilanes hydrolysis: An empirical approach. *Sci. Rep.* **2019**, *9*, 1–15. [CrossRef]

22. Echeverría, J.C.; Moriones, P.; Arzamendi, G.; Garrido, J.J.; Gil, M.J.; Cornejo, A.; Martínez-Merino, V. Kinetics of the acid-catalyzed hydrolysis of tetraethoxysilane (TEOS) by 29Si NMR spectroscopy and mathematical modeling. *J. Sol-Gel Sci. Technol.* **2018**, *86*, 316–328. [CrossRef]

23. Moriones, P.; Arzamendi, G.; Cornejo, A.; Garrido, J.J.; Echeverria, J.C. Comprehensive Kinetics of Hydrolysis of Organotriethoxysilanes by 29Si NMR. *J. Phys. Chem. A* **2019**, *123*, 10364–10371. [CrossRef]

24. Colin, B.; Lavastre, O.; Fouquay, S.; Michaud, G.; Simon, F.; Laferte, O.; Brusson, J.-M. Contactless Raman Spectroscopy-Based Monitoring of Physical States of Silyl-Modified Polymers during Cross-Linking. *Green Sustain. Chem.* **2016**, *6*, 151–166. [CrossRef]

25. Innocenzi, P.; Falcaro, P.; Grosso, D.; Babonneau, F. Microstructural evolution and order-disorder transitions in mesoporous silica films studied by FTIR spectroscopy. *Mater. Res. Soc. Symp. Proc.* **2002**, *726*, 271–281. [CrossRef]

26. Ponton, S.; Dhainaut, F.; Vergnes, H.; Samelor, D.; Sadowski, D.; Rouessac, V.; Lecoq, H.; Sauvage, T.; Caussat, B.; Vahlas, C. Investigation of the densification mechanisms and corrosion resistance of amorphous silica films. *J. Non Cryst. Solids* **2019**, *515*, 34–41. [CrossRef]

27. Fidalgo, A.; Ciriminna, R.; Ilharco, L.M.; Pagliaro, M. Role of the alkyl-alkoxide precursor on the structure and catalytic properties of hybrid sol-gel catalysts. *Chem. Mater.* **2005**, *17*, 6686–6694. [CrossRef]

28. Saputra, R.E.; Astuti, Y.; Darmawan, A. Hydrophobicity of silica thin films: The deconvolution and interpretation by Fourier-transform infrared spectroscopy. *Spectrochim. Acta Part A Mol. Biomol. Spectrosc.* **2018**, *199*, 12–20. [CrossRef]

29. Stocker, M.K.; Sanson, M.L.; Bernardes, A.A.; Netto, A.M.; Brambilla, R. Acid–base sensor based on sol–gel encapsulation of bromothymol blue in silica: Application for milk spoilage detection. *J. Sol-Gel Sci. Technol.* **2021**, *98*, 568–579. [CrossRef]

30. Fidalgo, A.; Ilharco, L.M. Correlation between physical properties and structure of silica xerogels. *J. Non Cryst. Solids* **2004**, *347*, 128–137. [CrossRef]

31. Izaak, T.I.; Martynova, D.O.; Stonkus, O.A.; Slavinskaya, E.M.; Boronin, A.I. Deposition of silver nanoparticles into porous system of sol-gel silica monoliths and properties of silver/porous silica composites. *J. Sol-Gel Sci. Technol.* **2013**, *68*, 471–478. [CrossRef]

32. Capeletti, L.B.; Zimnoch, J.H. Fourier Transform Infrared and Raman Characterization of Silica-Based Materials. In *Applications of Molecular Spectroscopy to Current Research in the Chemical and Biological Sciences*, 1st ed.; Stauffer, M., Ed.; Intechopen: Kansas City, KS, USA, 2016; pp. 3–22, ISBN 978-953-51-2681-2.

33. Echeverría, J.C.; Faustini, M.; Garrido, J.J. Effects of the porous texture and surface chemistry of silica xerogels on the sensitivity of fiber-optic sensors toward VOCs. *Sens. Actuators B Chem.* **2016**, *222*, 1166–1174. [CrossRef]

34. Echeverría, J.C.; Calleja, I.; Moriones, P.; Garrido, J.J. Fiber optic sensors based on hybrid phenyl-silica xerogel films to detect n-hexane: Determination of the isosteric enthalpy of adsorption. *Beilstein J. Nanotechnol.* **2017**, *8*, 475–484. [CrossRef] [PubMed]

35. Cruz-Quesada, G.; Espinal-Viguri, M.; Garrido, J. Novel Organochlorinated Xerogels: From Microporous Materials to Ordered Domains. *Polymers* **2021**, *13*, 1415. [CrossRef]

36. Rios, X.; Moriones, P.; Echeverría, J.C.; Luquín, A.; Laguna, M.; Garrido, J.J. Characterisation of hybrid xerogels synthesised in acid media using methyltriethoxysilane (MTEOS) and tetraethoxysilane (TEOS) as precursors. *Adsorption* **2011**, *17*, 583–593. [CrossRef]

37. Rios, X.; Moriones, P.; Echeverría, J.C.; Luquin, A.; Laguna, M.; Garrido, J.J. Ethyl group as matrix modifier and inducer of ordered domains in hybrid xerogels synthesised in acidic media using ethyltriethoxysilane (ETEOS) and tetraethoxysilane (TEOS) as precursors. *Mater. Chem. Phys.* **2013**, *141*, 166–174. [CrossRef]

38. Moriones, P. Sintesis y Caracterización de Xerogeles Silíceos Híbridos (RTEOS/TEOS.; R = P, Ph). Ph.D. Thesis, Universidad Publica de Navarra, Pamplona, Spain, 2015. Available online: https://academica-e.unavarra.es/handle/2454/20351 (accessed on 14 November 2020).

39. Torres-Luna, J.A.; Carriazo, J.G. Porous aluminosilicic solids obtained by thermal-acid modification of a commercial kaolinite-type natural clay. *Solid State Sci.* **2019**, *88*, 29–35. [CrossRef]

40. Sakka, S. Handbook of Sol-Gel Science and technology processing characterization and applications. In *Characterization of Sol-Gel materials and Products*; Kluwer Academic Publishers: Amsterdam, The Netherlands, 2005; Volume II, ISBN 1-4020-7967-2.

41. Berrier, E.; Courtheoux, L.; Bouazaoui, M.; Capoen, B.; Turrell, S. Correlation between gelation time, structure and texture of low-doped silica gels. *Phys. Chem. Chem. Phys.* **2010**, *12*, 14477–14484. [CrossRef]

42. Salon, B.M.C.; Belgacem, M.N. Competition between hydrolysis and condensation reactions of trialkoxysilanes, as a function of the amount of water and the nature of the organic group. *Colloids Surf. A Physicochem. Eng. Asp.* **2010**, *366*, 147–154. [CrossRef]

43. Pierre, A.C. *Introduction to Sol-Gel Processing*; Springer: Berlin/Heidelberg, Germany, 2020; ISBN 9780792381211.

44. Benbow, J.J. *Ultrastructure Processing of Advance Materials.*; Uhlmann, D.R., Ulrich, D.R., Eds.; John Wiley & Sons, INC.: New York, NY, USA, 1989; Volume 44, ISBN 0471529869.

45. Depla, A.; Verheyen, E.; Veyfeyken, A.; Van Houteghem, M.; Houthoofd, K.; Van Speybroeck, V.; Waroquier, M.; Kirschhock, C.E.A.; Martens, J.A. UV-Raman and 29Si NMR spectroscopy investigation of the nature of silicate oligomers formed by acid catalyzed hydrolysis and polycondensation of tetramethylorthosilicate. *J. Phys. Chem. C* **2011**, *115*, 11077–11088. [CrossRef]

46. Depla, A.; Lesthaeghe, D.; Van Erp, T.S.; Aerts, A.; Houthoofd, K.; Fan, F.; Li, C.; Van Speybroeck, V.; Waroquier, M.; Kirschhock, C.E.A.; et al. 29Si NMR and UV-raman investigation of initial oligomerization reaction pathways in acid-catalyzed silica Sol-Gel chemistry. *J. Phys. Chem. C* **2011**, *115*, 3562–3571. [CrossRef]

47. Fidalgo, A.; Ilharco, L.M. Chemical Tailoring of Porous Silica Xerogels: Local Structure by Vibrational Spectroscopy. *Chem. A Eur. J.* **2004**, *10*, 392–398. [CrossRef]

48. Hayami, R.; Ideno, Y.; Sato, Y.; Tsukagoshi, H.; Yamamoto, K.; Gunji, T. Soluble ethane-bridged silsesquioxane polymer by hydrolysis–condensation of bis(trimethoxysilyl)ethane: Characterization and mixing in organic polymers. *J. Polym. Res.* **2020**, *27*, 1–10. [CrossRef]

49. Ramezani, M.; Vaezi, M.R.; Kazemzadeh, A. The influence of the hydrophobic agent, catalyst, solvent and water content on the wetting properties of the silica films prepared by one-step sol-gel method. *Appl. Surf. Sci.* **2015**, *326*, 99–106. [CrossRef]

50. Innocenzi, P. Infrared spectroscopy of sol–gel derived silica-based films: A spectra-microstructure overview. *J. Non Cryst. Solids* **2003**, *316*, 309–319. [CrossRef]

51. Handke, M.; Kowalewska, A. Siloxane and silsesquioxane molecules—Precursors for silicate materials. *Spectrochim. Acta Part. A Mol. Biomol. Spectrosc.* **2011**, *79*, 749–757. [CrossRef]

52. Coates, J. Interpretation of Infrared Spectra, A Practical Approach. In *Encyclopedia of Analytical Chemistry*, 3rd ed.; Meyers, R.A., Ed.; Wiley & Sons, Ltd.: New York, NY, USA, 2006; pp. 1–23, ISBN 978-0-471-97670-7.

53. Launer, P.J.; Arkles, B. Infrared Analysis of Orgaonsilicon Compounds. In *Silicon Compounds: Silanes and Silicones*, 3rd ed.; Gelest Inc.: Morrisville, PA, USA, 2013; pp. 175–178.

54. Chen, G.; Zhou, Y.; Wang, X.; Li, J.; Xue, S.; Liu, Y.; Wang, Q.; Wang, J. Construction of porous cationic frameworks by crosslinking polyhedral oligomeric silsesquioxane units with N-heterocyclic linkers. *Sci. Rep.* **2015**, *5*, 1–14. [CrossRef]

55. Park, E.S.; Ro, H.W.; Nguyen, C.V.; Jaffe, R.L.; Yoon, D.Y. Infrared spectroscopy study of microstructures of poly(silsesquioxane)s. *Chem. Mater.* **2008**, *20*, 1548–1554. [CrossRef]

56. Kamiya, K.; Dohkai, T.; Wada, M.; Hashimoto, T.; Matsuoka, J.; Nasu, H. X-ray diffraction of silica gels made by sol-gel method under different conditions. *J. Non Cryst. Solids* **1998**, *240*, 202–211. [CrossRef]

57. García-Cerda, L.A.; Mendoza-González, O.; Pérez-Robles, J.F.; González-Hernández, J. Structural characterization and properties of colloidal silica coatings on copper substrates. *Mater. Lett.* **2002**, *56*, 450–453. [CrossRef]

58. Hagiwara, Y.; Shimojima, A.; Kuroda, K. Alkoxysilylated-derivatives of double-four-ring silicate as novel building blocks of silica-based materials. *Chem. Mater.* **2008**, *20*, 1147–1153. [CrossRef]

59. Nowacka, M.; Kowalewska, A.; Makowski, T. Structural studies on ladder phenylsilsesquioxane oligomers formed by polycondensation of cyclotetrasiloxanetetraols. *Polymer* **2016**, *87*, 81–89. [CrossRef]

60. Ospino, I.; Luquin, A.; Jiménez-Ruiz, M.; Pérez-Landazábal, J.I.; Recarte, V.; Echeverría, J.C.; Laguna, M.; Urtasun, A.A.; Garrido, J.J. Computational Modeling and Inelastic Neutron Scattering Contributions to the Study of Methyl-silica Xerogels: A Combined Theoretical and Experimental Analysis. *J. Phys. Chem. C* **2017**, *121*, 22836–22845. [CrossRef]

61. Handke, M.; Jastrzębski, W. Vibrational spectroscopy of the ring structures in silicates and siloxanes. *J. Mol. Struct.* **2004**, *704*, 63–69. [CrossRef]

62. Shi, Y.; Neuefeind, J.; Ma, D.; Page, K.; Lamberson, L.A.; Smith, N.J.; Tandia, A.; Song, A.P. Ring size distribution in silicate glasses revealed by neutron scattering first sharp diffraction peak analysis. *J. Non Cryst. Solids* **2019**, *516*, 71–81. [CrossRef]

63. Tan, C.Z.; Arndt, J. Interaction of longitudinal and transverse optic modes in silica glass. *J. Chem. Phys.* **2000**, *112*, 5970–5974. [CrossRef]
64. Caresani, J.R.; Lattuada, R.M.; Radtke, C.; Dos Santos, J.H.Z. Attempts made to heterogenize MAO via encapsulation within silica through a non-hydrolytic sol-gel process. *Powder Technol.* **2014**, *252*, 56–64. [CrossRef]
65. Choi, S.S.; Lee, A.S.; Lee, H.S.; Baek, K.Y.; Choi, D.H.; Hwang, S.S. Synthesis and characterization of ladder-like structured polysilsesquioxane with carbazole group. *Macromol. Res.* **2011**, *19*, 261–265. [CrossRef]
66. Gallardo, J.; Durán, A.; Di Martino, D.; Almeida, R.M. Structure of inorganic and hybrid SiO2 sol-gel coatings studied by variable incidence infrared spectroscopy. *J. Non Cryst. Solids* **2002**, *298*, 219–225. [CrossRef]

 *polymers*

Article

Synthesis, Optical and Electrical Characterization of Amino-alcohol Based Sol-gel Hybrid Materials

Bárbara R. Gomes, Rita B. Figueira *, Susana P. G. Costa, M. Manuela M. Raposo and Carlos J. R. Silva †

Centro de Química, Universidade do Minho, Campus de Gualtar, 4710-057 Braga, Portugal;
barbara.sgomes11@gmail.com (B.R.G.); spc@quimica.uminho.pt (S.P.G.C.);
mfox@quimica.uminho.pt (M.M.M.R.)
* Correspondence: rbacelarfigueira@quimica.uminho.pt or rita@figueira.pt
† This paper is dedicated to the memory of Professor C. J. R. Silva who passed away suddenly on 27 August 2020.

Received: 22 October 2020; Accepted: 10 November 2020; Published: 12 November 2020

Abstract: This manuscript describes the synthesis and characterization of five new organic–inorganic hybrid (OIH) sol-gel materials that were obtained from a functionalized siloxane 3-glycidoxypropyltrimethoxysilane (GPTMS) by the reaction with the new Jeffamine®, namely three different diamines, i.e., EDR-148, RFD-270, and THF-170, a secondary diamine, i.e., SD-2001, and a triamine, i.e., T-403. The OIH sol-gel materials were characterized by UV-visible absorption spectrophotometry, steady-state photoluminescence spectroscopy, and electrochemical impedance spectroscopy. The reported OIH sol-gel materials showed that, with the exception of the samples prepared with Jeffamine® SD-2001, the transmittance values ranged between 61% and 79%. Regarding the capacitance data, the values reported changed between 0.008 and 0.013 nF cm^{-2}. Due to their optical and electrical properties these new OIH materials show promising properties for applications as support films in an optical sensor area such as fiber sensor devices. Studies to assess the chemical stability of the OIH materials in contact with cement pastes after 7, 14, and 28 days were also performed. The samples prepared with THF–170 and GPTMS, when compared to the samples prepared with RFD-270 and T-403, exhibited improved behavior in the cement paste (alkaline environment), showing promising properties for application as support film in optical fiber sensors in the civil engineering field.

Keywords: hybrids; sol-gel; materials; amino-alcohol

1. Introduction

In the last few decades, the design and development of organic–inorganic hybrid (OIH) sol-gel materials for a wide range of applications has achieved a high scientific proficiency level. Several OIH materials have been reported for applications such as coatings for corrosion mitigation, smart windows, photochromic and electrochromic materials, sensors, optical filters, and absorbers, among others. The sol-gel method is recognized as a green, low-cost, and versatile route that allows organic–inorganic hybrid (OIH) materials to be obtained in a simple way by the reaction between polyetheramines and organoalkoxysilanes. The most remarkable properties of such OIH materials include chemical stability, suitable dielectric properties, selective ion binding, and ion conducting ability, depending on their organic and inorganic components. Such materials have also been reported as interesting luminescent sources [1–5] and hosts of luminescent species [6–13]. For instance, Severo Rodembusch et al. reported for the first time the synthesis of fluorescent hybrid aerogels in the blue–green–yellow region. The dyes were obtained by the reaction of amino benzazole derivatives with 3-(triethoxysilyl)propyl isocyanate.

As precursor a pre-polymerized tetraethoxysilane was used [8]. In 2015, Brito et al. reported the entrapment of blue–green luminescent C-dots in transparent silica. The silica was obtained by pyrolysis of the methyl groups present in the nanometric silica grains [12].

Other materials such as siloxane-polyether OIH materials, also known as di- or tri-ureasil, have also been widely reported [11,14–26] for an extensive range of applications. The most common strategy to prepare such materials with the sol-gel method involves several hydrolysis and polycondensation reactions allowing their morphology, structure, and chemical composition to be tuned. For instance, Boev et al. [15] reported a simple procedure for the synthesis of ureasilicate materials. The authors studied the influence of the molar ratio between 3-isocyanate propyltriethoxysilane (ICPTES) and Jeffamine® ED-600. A homogeneous and flexible material was attained and it was concluded that by changing the catalyst nature and the molar ratio of ICPTES/Jeffamine® highly transparent samples within the visible range with dissimilar elastic properties were obtained. Later, the studied materials were doped with different nanoparticles, including CdS [16,26–28], CdSe [20,29], $Zn_xCd_{1-x}S$ [17], and PbS [30]. Ureasilicate materials have also been reported as coatings for corrosion mitigation in alkaline environments [19,22,31] and in the presence of chloride ions [32,33], showing good performance for the desired purpose. These materials were also reported as films for the development of a relative humidity (RH) optical fiber sensor [23]. The ureasilicate material was deposited by dip coating the fiber and studied in the range of 5% to 95% of RH. It was also concluded that when compared to other polymer-based solutions, the proposed material showed enhanced durability and sensitivity (22.2 pm/% RH) in monitoring the RH of two concrete blocks for one year [23]. Willis-Fox et al. [34] reported a strategy for the preparation of a conjugated polymer-di-ureasil composite material showing a tunable emission color from blue to yellow using a simple sol-gel method. Moreira et al. [18] reported the synthesis of ureasilicates and amino-alcohol silicate OIH sol-gel materials. The amino-alcohol silicates were obtained by reacting Jeffamine® ED-600 or Jeffamine® ED-900 with 3-glycidoxypropyltrimethoxysilane (GPTMS) using a molar ratio of GPTMS:Jeffamine® = 2:1. The GPTMS precursor is a suitable molecule establishing amino-alcohol bonds between the inorganic (silicate) and organic (Jeffamine) components (i.e., a polyether chain) similar to that reported for the ureasilicate matrix [18]. The authors obtained flexible xerogels with high transparency within the visible range and showed that the amino-alcohol silicate samples exhibited suitable optical and mechanical properties that can lead to the production of low-cost and variable shape diffraction lenses in a wide range of substrates for optical applications [18]. Figueira et al. also reported the use of amino-alcohol silicate matrices as coating materials for corrosion mitigation of galvanized steel in an alkaline environment [22,35].

In the last few decades the use of optical fiber sensors (OFS) to monitor physical parameters of structures in the civil engineering field has been widely employed. For instance, the application of OFS to monitor properties such as the curvature and deflection of bridges [36–39], the onset of cracking [40–42], and temperature [43,44] has been consistently reported. Nevertheless, besides physical properties, other parameters such as pH level [45,46], relative humidity [47–49], chloride content [50–52], and alkali-silica reactions (ASR) [53] are also responsible for earlier failure of civil engineering structures. The development of robust and accurate monitoring sensing systems is of extreme importance as they will allow the service life of concrete structures to increase and costs saved at the same time. The monitoring of concrete structures using functionalized optical fiber sensors (OFS) with sol-gel films revealed to be an interesting approach to increase its service life [54].

Considering the steadily increasing attention being paid to these OIH films together with their promising properties, further research is needed. Therefore, this work reports the synthesis and optical and electrochemical characterization of new OIH amino-alcohol silicate-based materials for potential application on an OFS for health monitoring of concrete structures. In this manuscript, the new Jeffamines® will be considered. To the authors' knowledge no studies using the Jeffamine® EDR-148, Jeffamine® RFD-270, Jeffamine® THF-170, Jeffamine® SD-2001, and Jeffamine® T-403 have been reported. Therefore, three different diamines (Jeffamine® EDR-148, Jeffamine® RFD-270, and Jeffamine® THF-170), a secondary diamine (Jeffamine® SD-2001), and a triamine (Jeffamine® T-403) were made to react with GPTMS-producing amino-alcohol silicate matrices. More specifically, five new different OIH sol-gel materials were synthesized.

To the authors' knowledge no studies have been reported on the synthesis of OIH xerogel films using GPTMS and Jeffamine® EDR-148, Jeffamine® RFD-270, or Jeffamine® THF-170 as a siloxane precursor. Concerning Jeffamine® T-403, this precursor has been reported by different authors for a wide range of applications [55–59]. For instance, Spírkpvá et al. in 2003 reported the synthesis and structural and mechanical properties of the OIHs obtained by the reaction between GPTMS and three different Jeffamines®: T-403, D-230, and D-400 [55]. Nevertheless, the authors used different ratios and synthesis parameters [55].

The OIH sol-gel materials synthesized were characterized by UV–visible spectrophotometry, steady-state photoluminescence spectroscopy, and electrochemical impedance spectroscopy. The reported OIH sol-gel materials show interesting optical properties for application as a support matrix in OFS and the capacitance values changed between 0.008 and 0.013 nF cm^{-2}. Preliminary studies were also conducted to assess the chemical stability of the OIH materials in contact with cement pastes for 7, 14, and 28 days. The samples prepared with Jeffamine® THF-170 and GPTMS, when compared to the others, exhibited improved and promising properties in a cement paste environment.

2. Experimental

2.1. Reagents

Figure 1 shows the structure and acronym (in bold) of the precursors used in the synthesis of the OIH sol-gel materials. The structure of Jeffamine® THF-170 (*Poly(oxy-1,4-butanediyl), α-hydro-Ω-hydroxy-,polymer with ammonia*) (Huntsman Corporation, Pamplona, Spain), hereafter referred to as THF-170, was not provided by the supplier. The structure which is indicated in Figure 1 is based on the information reported by Létoffé et al. for a similar Jeffamine® (Jeffamine® THF-100) [60]. Five different Jeffamines® (structurally identified as polyetheramines) were kindly supplied by Huntsman, with different molecular weights and reactivity. Jeffamine® EDR-148 is the most reactive Jeffamine® when compared to diamines and triamines due to the unhindered nature of the amine groups (vide Figure 1).

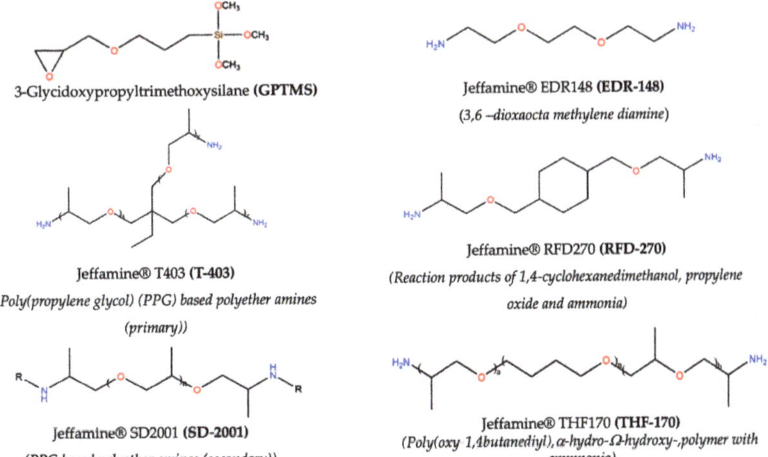

Figure 1. Structure and acronym (in bold) of the precursors used in the synthesis of the organic–inorganic hybrid sol-gel materials.

Jeffamine® RFD-270 is an amine containing both rigid (cycloaliphatic) and flexible (etheramine) moieties in the same molecule and the acronym RFD stands for "rigiflex diamine" (vide Figure 1). Jeffamine® T-403 is a triamine prepared by the reaction of propylene oxide with a triol initiator, followed by the amination of the terminal hydroxyl groups (vide Figure 1). The Jeffamine® THF-170 is based on a poly(tetramethylene ether glycol)]/(poly(propylene glycol) copolymer. Jeffamine® SD-2001 is a polyetheramine, where SD stands for "secondary diamine." The secondary amine groups provide a much slower reaction compared to primary amine groups.

All the precursors used, namely the Jeffamines® and the GPTMS (97%, Sigma-Aldrich, St. Louis, MO, USA), were used as supplied. As solvents, tetrahydrofuran (99.5% stabilized with ~ 300 ppm of BHT, Panreac, Darmstadt, Germany) and ultra-pure water with high resistivity (higher than 18 MΩ cm) obtained from a Millipore water purification system (Milli-Q®, Merck KGaA, Darmstadt, Germany) were employed.

2.2. Synthesis Procedure of Xerogel Films

The obtained xerogel films were synthesized in sequential steps, as schematized in Scheme 1.

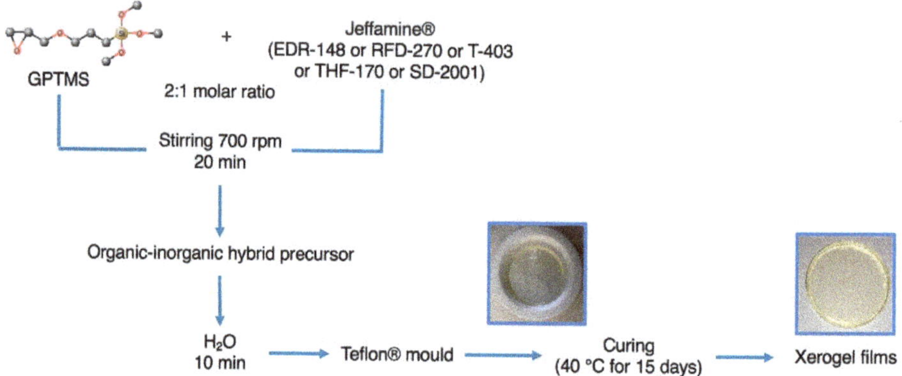

Scheme 1. Main steps involved in the synthesis of the OIH sol-gel matrices.

Before the 3-glycidoxypropyltrimethoxysilane (GPTMS) addition, the Jeffamines® were solubilized using tetrahydrofuran. After 10 min, the GPTMS was added to the Jeffamine® (EDR-148, RFD-270, T-403, THF-170, or SD-2001) using a molar ratio of 2 GPTMS:1 Jeffamine® in a glass vessel under stirring at 700 rpm for 20 min, with the exception of Jeffamine® RFD-270, which was prepared with a molar ratio of 1 GPTMS:1 Jeffamine®.

The reaction between the amine end group (–NH$_2$) of the Jeffamine® with different molecular weight and structure (Jeffamine® EDR-148, Jeffamine® RFD-270, Jeffamine® T-403, Jeffamine® THF-170, and Jeffamine® SD-2001) and the epoxy group of GPTMS led to the formation of the OIH precursors of the future gel matrices that are named as conventional amino-alcohol silicates and referred to as A(148), A(270), A(403), A(170), and A(2001). The numbers in parentheses stand for the Jeffamine® used, with the reference names indicated in Figure 1. The second step included the addition of H$_2$O to obtain an H$_2$O:Jeffamine® molar ratio equal to 5.94. The mixture was stirred for another 10 min and placed into a Teflon® mold (DuPont, Wilmington, DE, USA) and sealed with Parafilm® that was pin-holed and placed in an oven (UNB 200, Memmert, Buechenbach, Germany) and kept at 40 °C for 15 days. This procedure ensures precise control and reproducibility of the conditions of hydrolysis/condensation reactions as well as the evaporation of the residual solvents. Scheme 1 shows that the implemented preparation conditions applied allowed homogeneous and transparent film samples free of cracks to be obtained.

2.3. OIH Sol-Gel Film Characterization

The OIH materials synthesized were characterized optically and electrochemically. All the measurements were conducted at room temperature. The UV–visible transmission and absorption spectra of the OIH film samples were obtained using a spectrophotometer (UV-2501 PC, Shimadzu, Duisburg, Germany) in the range of 250–700 nm. The fluorescence spectra were obtained using a spectrofluorometer (Fluoromax–4, Horiba Jovin Yvon, Madrid, Spain). The emission spectra were recorded in the wavelength range of 300–700 nm using different excitation wavelengths and acquired at front-face geometry at room temperature.

Electrochemical impedance spectroscopy (EIS) measurements were carried out to characterize resistance, electrical conductivity, and the electric permittivity of the prepared OIH films. Measurements were performed using a two-electrode system in which the film was placed between two parallel Au disc electrodes (10 mm diameter and 250 μm thickness) and a support cell [33]. Measurements were performed in a Faraday cage using a potentiostat/galvanostat/ZRA (Reference 600+, Gamry Instruments, Warminster, PA, USA) by applying a 10 mV (peak-to-peak, sinusoidal) electrical potential within a frequency range from 1×10^5 Hz to 0.01 Hz (10 points per decade) at open circuit potential. The frequency response data were displayed in a Nyquist plot. The Gamry ESA410 Data Acquisition software was used for data fitting purposes.

2.4. Preparation of Cement Paste

The chemical stability of the different OIH films was studied in a cement paste. The cement paste was prepared using cement type I 42,5R and distilled water. The ratio of water to cement (w/c) used was equal to 0.5. The OIH films used were 10 mm in diameter and the thickness ranged between 1.4 and 1.8 mm.

3. Results and Discussion

3.1. UV-Visible Spectrophotometry Analysis

Figure 2 shows the optical transmittance spectra as a function of wavelength obtained for the synthesized xerogel films A(148), A(270), A(403), A(170), and A(2001). The A(2001) samples show the lowest transmittance, followed by A(148), A(170), A(270), and A(403), with A(403) showing the highest transmittance values of between 400 and 700 nm.

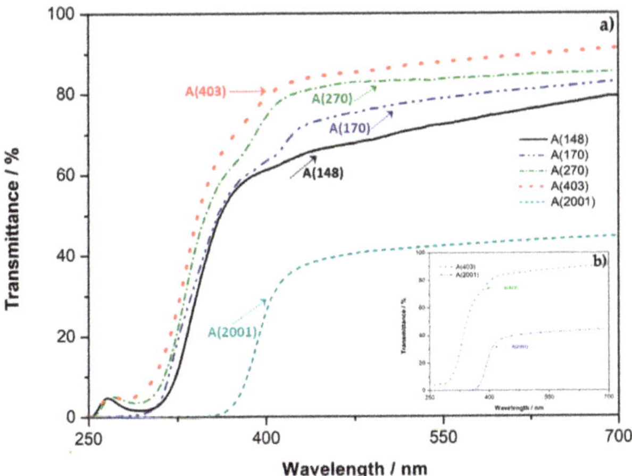

Figure 2. UV-visible transmission spectra obtained for xerogel films (**a**) A(148), A(270), A(403), A(170), and A(2001) and as an inset detail of (**b**) A(403) and A(2001).

In the UV region between 250 and 300 nm, all the samples showed low transmittance, which is in accordance with the literature [18]. The transmittance data shows that at 400 nm wavelength, the xerogel that provided the highest value was A(403) with around 79%. At the same wavelength, A(270) showed a transmittance of around 74% and the lowest transmittance was given by A(2001) with a value of around 26%. At 400 nm, A(148) and A(170) showed a transmittance of 61% and 63%, respectively. As the wavelength increases the transmittance values increase, with A(403) reaching the highest value, i.e., 89% at 600 nm.

These results are in agreement with the results reported by Moreira et al. [18] and by Erdem et al. [61]. Both authors reported that the OIHs obtained with lower molecular weights show higher transmittance than films prepared with higher molecular weights. Moreira et al. [18] described the reaction between Jeffamine® ED-600 and GPTMS as showing higher transmittance than films prepared by the reaction between Jeffamine® ED-900 and GPTMS. This behavior suggests that the OIH transmittance is related to the molecular weight of the Jeffamine®. Another explanation may be related to the structure involved together with the interactions established between the Jeffamine® and the GPTMS. Zea Bermudez et al. reported a study focused on the synthesis and FTIR characterization of OIHs using as precursors Jeffamines® ED-600, ED-900, and ED-2000 with 3-isocyanatepropyltriethoxysilane. It was reported that the FTIR spectrum of U(2000) showed that the polyether chains of the parent diamine become less ordered upon incorporation into the inorganic backbone. The number of oxyethylene units present affected the amide I and amide II bands. This indicated that the N–H groups of the urea linkage were involved in the hydrogen bonds of different strengths. Moreover, the authors proposed the existence of non-hydrogen-bonded urea groups and hydrogen-bonded urea–urea and urea–polyether associations. On one hand the formation of urea–urea structures was favored in the U(600), while on the other hand the number of free carbonyl groups was higher in U(2000) [14]. Similar behavior for the A(2001) may be expected, which may justify the huge difference of transmittance between A(148), A(170), A(270), and A(403) since Jeffamine® SD-2001 is a difunctional secondary amine derived from Jeffamine® ED-2000. Nevertheless, since in this case GPTMS was used instead of 3-isocyanatepropyltriethoxysilane no further conclusions can be drawn. Therefore, further studies, namely FTIR analysis, should be conducted in order to clarify the interactions between the precursors.

The thickness of the films ranged between 0.913 mm and 1.434 mm. The thickness obtained for the OIHs for A(148), A(270), A(403), A(170), and A(2001) was 1.301 mm, 0.913 mm, 1.051 mm, 1.434 mm, and 1.024 mm, respectively. Therefore, the absorbance measurements were divided by the thickness obtained for each film. Figure 3 shows the UV-vis absorption spectra normalized to the film thickness of the four synthesized xerogel films (A(148), A(270), A(403), A(170), and A(2001)).

The highest absorbance peaks were found in the UV region between 250 and 300 nm (Figure 3). The maximum absorption wavelength obtained for each xerogel film for A(148), A(270), A(403), A(170), and A(2001) was 294 nm, 290 nm, 281 nm, 276 nm, and 310 nm, respectively. Regardless of the structure of the Jeffamine® used, as the molecular weight of Jeffamine® increases, the maximum wavelength in the UV region (between 250–300 nm) decreases. An exception was found for A(2001), which showed the highest molecular weight and a maximum at a wavelength of around 310 nm. The highest and the lowest absorbance peaks were obtained for A(2001) and A(403) xerogel films. Figure 3 also shows that, with the exception of A(2001), all the other OIH matrices are transparent from 300 nm forward. Therefore, these materials are suitable for probe immobilization in the range between 300 and 700 nm. Considering the low transmittance obtained for A(2001) and its translucency, no further studies were conducted in characterizing A(2001). Moreover, previous studies showed that as the molecular weight of a Jeffamine® increases, the resistance of the OIHs to a highly alkaline environment decreases [22,32].

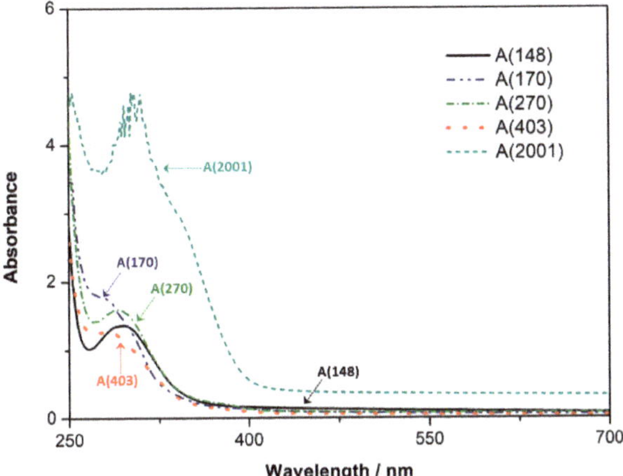

Figure 3. UV-visible absorption spectra normalized for the thickness of each xerogel film, namely A(148), A(270), A(403), A(170), and A(2001).

3.2. Photoluminescence Spectrophotometry Analysis

Figure 4 shows the emission fluorescence spectra obtained for the synthesized xerogel films A(148), A(270), A(403), and A(170) at different excitation wavelengths. The limits of excitation wavelengths were defined according to the maximum wavelength recorded for each OIH film in UV-vis analysis, always starting at 250 nm.

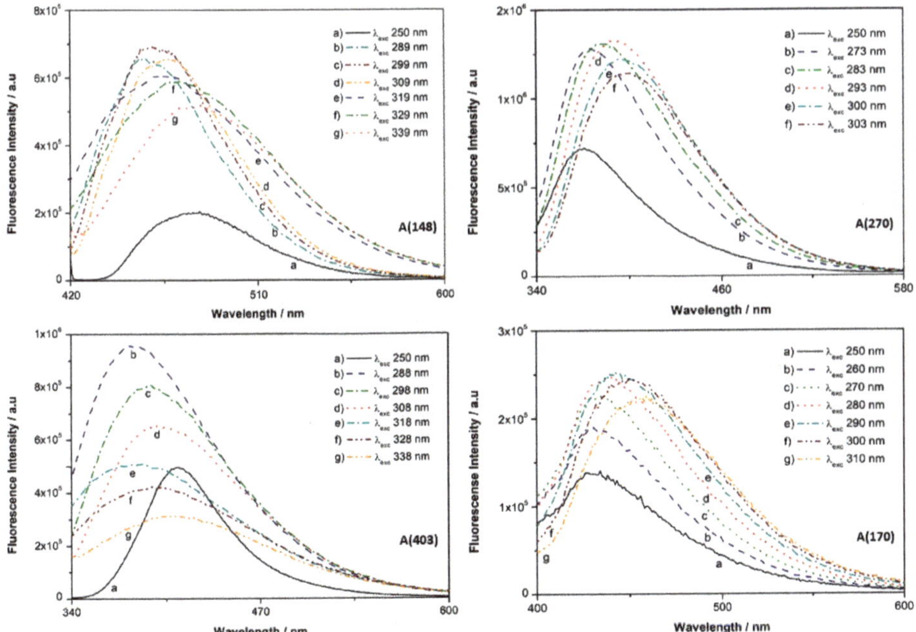

Figure 4. Photoluminescence spectra obtained for xerogel films A(148), A(270), A(403), and A(170).

Figure 4 shows that the OIH materials show an intrinsic emission. In the case of the A(270) and A(170) OIH matrices, it can be observed that the wavelength of the emission peak displaces to higher wavelengths when the excitation wavelength increases. This intrinsic emission linked to these OIH matrices is due to the photoinduced proton transfer between defects NH_3^+/NH^- and due to the electron–hole recombination occurring in the siloxane nanoclusters [62]. The emission wavelength dependency with the excitation energy is connected to unorganized processes that are generally linked to transitions that occur between localized states in non-crystalline structures [62]. According to Carlos et al. [2], in OIHs similar to the ones reported here the hierarchy in the silica backbone dimension defines the emission wavelength. This is in agreement with the findings reported here. This energy dependence of the emission wavelength with the excitation energy is related to the size of the silica clusters [63]. The same authors reported that larger clusters emit at longer wavelengths than smaller clusters. Generally, shorter polyether chains (A(148) and A(403)) provide samples with higher photoluminescence while larger chains may induce a dilution effect that may reduce the luminescence efficiency [63], which is in agreement with the data reported here. In the case of A(148) and A(403) it can be observed that the wavelength of the emission peak displaces to lower wavelengths when the excitation wavelength increases. Photoluminescence spectra of the samples with shorter polymer chains such as the ones used in A(148) and A(403) suggest that the electron–hole recombination occurring in the siloxane nanoclusters is large enough to allow efficient energy-transfer mechanisms.

The full width at half maximum did not decrease with decreasing excitation energy. Considering that the OIH materials are a biphasic system (i.e., organic and inorganic components are mixed at nanometric scale), this may contribute to a higher light scattering that may lead to changes within the sample and to a general broadening of the peaks. Generally, Figure 4 shows that the full width at half maximum did not decrease with decreasing excitation energy.

Figure 5 shows the photographs of the xerogel samples synthesized and their photoluminescence response when excited with UV light, i.e., an excitation wavelength of 365 nm.

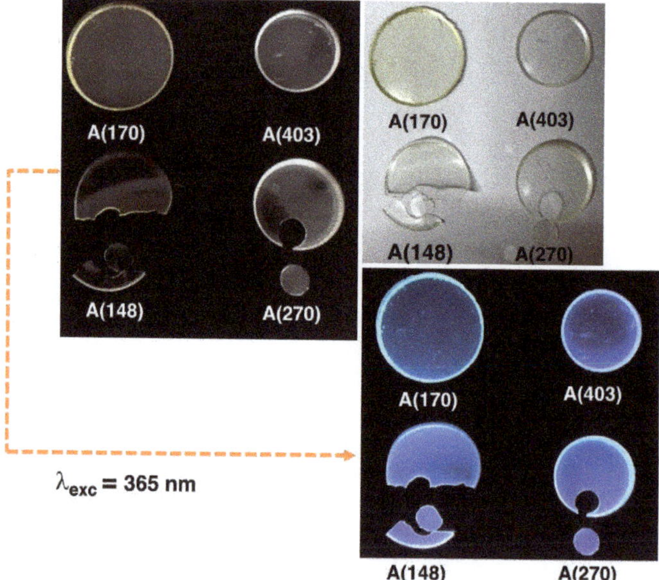

Figure 5. Amino-alcohol silicate samples and their photoluminescence response after an excitation with a light with a wavelength of 365 nm.

The blue color of the A(148), A(270), A(403), and A(170) xerogels after excitation with UV radiation arises from the photoluminescence emission in the 450–470 nm region [29]. It is generally accepted that silicon-based materials can emit light within a wide energy region that ranges from ultraviolet to infrared [2,64,65]. The backbone silicon-based structures are responsible for the emission energy, i.e., an increase in the siliceous network may result in a decrease in the corresponding energy gap [2]. The emission energy of silicon-based materials depends on the hierarchy of their backbone dimensions. Since the network dimension changes, the band gap energies change accordingly. This dependence of the energy gap on the backbone dimensions is related to the extension of the silicon σ-conjugations through the OIH network. These induce the delocalization of the electrons, leading to the formation of electronic band structures. Increasing their extension along the silicon backbone induces a decrease in the energy gap values, with a corresponding increase in the skeleton dimensions. Moreover, the Si–O–Si network is considered responsible for the blue emission of oxidized porous silicon [2].

3.3. EIS Measurements

Electrochemical impedance spectroscopy (EIS) is a very interesting technique widely used for the characterization of OIH sol-gel materials [32,33,35,66,67]. Figure 6 shows the Nyquist plots of the pure OIH films based on A(270) (Figure 6a), A(403) (Figure 6b), A(2001) (Figure 6c), and A(170) (Figure 6d) matrices, correspondingly. The experimental and the fitting results are shown in the Nyquist plots (Figure 6) and are schematized by squares and a continuous line, respectively. The equivalent electrical circuits (EEC) used for each sample were introduced as an inset in each Nyquist plot.

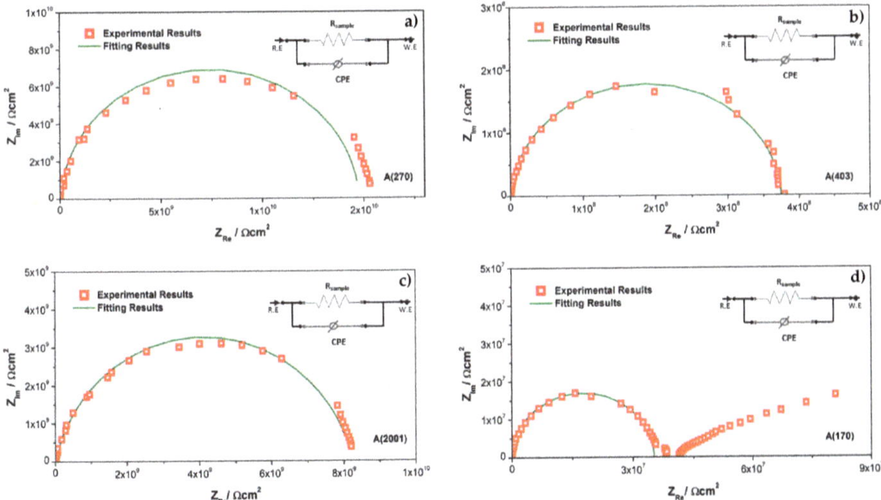

Figure 6. Complex plane impedance plots obtained for the OIH films based on: (**a**) A(270); (**b**) A(403); (**c**) A(2001); and (**d**) A(170) matrices (inset displayed in the EEC (equivalent electrical circuit) diagram adapted for fitting purposes).

The Nyquist plots illustrated in Figure 6 show that generally in the region of high frequencies a semicircle intersects the x-axis. It can also be observed that the diameter of the semicircle changes with the Jeffamine® used. This indicates that the dielectric properties of the OIH sol-gel materials (e.g., conductivity, capacitance, etc.) change with the different Jeffamine® used in the OIH synthesis. The data obtained at lower frequencies (Figure 6d) describes a line suggesting another electrochemical process that can be assigned to the interfacial phenomena between the gold electrodes and the OIH disc, which is in accordance with the literature [35]. However, this particular phenomenon will not be considered in the EIS analysis since it is not necessary to characterize the dielectric properties of the materials that are being studied. The EIS results regarding OIH A(148) films are not reported since it was not possible to perform its measurements. This may be due to the high rigidity and the difficulty in establishing the contact between the gold electrodes and the A(148) OIH discs. The high rigidity contributes to a poor electric contact between the OIH film and the gold electrode discs, disabling the electrical response.

The Nyquist plots (Figure 6) show a depressed form, therefore the analysis of all the impedance responses was based on an EEC where constant phase elements (CPE) were used in place of pure capacitance.

As reported, the impedance of a CPE can be defined as [68]:

$$Z_{CPE} = \frac{1}{[Q(j\omega)^\alpha]} \tag{1}$$

In Equation (1), both Q and α parameters are independent of the frequency. When $\alpha = 1$, Q stands for the capacity of the interface [67]. However, when $0 < \alpha < 1$ the system shows behavior that is generally linked to surface heterogeneity and the impedance for the EEC is given by Equation (2) [68]:

$$Z_{CPE} = \frac{R_{sample}}{\left[1 + (j\omega)^\alpha QR_{sample}\right]} \tag{2}$$

The interfacial capacitance (C_{eff}) is given by Equation (3) using the estimated Q value [68]:

$$C_{eff} = \left[QR_{sample}{}^{(1-\alpha)}\right]^{\frac{1}{\alpha}} \tag{3}$$

Table 1 shows the values of the proposed EEC elements obtained from the EIS data fitting for all the OIH materials synthesized.

Table 1. Values of elements of the proposed EEC obtained from EIS data fitting of the OIH materials synthesized and errors in percentage. For comparison purposes the data reported for A(400) [35], A(900) [35], and A(2000) [35] were also included.

OIH Sample	R_{Sample} (Ω)	CPE (Q) ($S^\alpha \, \Omega^{-1}$)	α	χ^2
A(270)	4.58×10^{10} ($\pm$ 0.83 %)	$6.75 \times 10^{-12.}$ ($\pm$ 1.63 %)	0.96	3.31×10^{-3}
A(400) [35]	8.53×10^{8} ($\pm$ 0.60 %) [35]	7.38×10^{-11} ($\pm$ 3.65%) [35]	0.73 [35]	NR
A(403)	4.67×10^{8} ($\pm$ 0.80 %)	$8.23 \times 10^{-12.}$ ($\pm$ 2.96 %)	0.97	4.08×10^{-4}
A(900) [35]	5.79×10^{4} ($\pm$ 0.19 %) [35]	4.20×10^{-10} ($\pm$ 4.34%) [35]	0.81 [35]	NR
A(170)	3.53×10^{7} ($\pm$ 0.77 %)	4.76×10^{-12} ($\pm$ 2.29 %)	0.98	3.20×10^{-4}
A(2000) [35]	3.62×10^{4} ($\pm$ 0.24%)	2.27×10^{-9} ($\pm$ 5.09%) [35]	0.71	NR
A(2001)	7.54×10^{9} ($\pm$ 0.853 %)	$1.50 \times 10^{-11.}$ ($\pm$ 1.71 %)	0.86	9.47×10^{-3}

NR—not reported; CPE—constant phase element.

Table 1 shows that all the resistances of the new OIH film samples synthesized, namely A(170), A(270), A(403), and A(2001), are between 10^7 and 10^{10} Ω. With the exception of the values obtained for A(400) [35] (vide Table 1), all the new synthesized OIH materials show higher values when compared to the ones already reported for A(900) and A(2000).

The values obtained for the elements of the EEC proposed and reported in Table 1, i.e., the values of resistance (R_{Sample}), the constant phase element (CPE), and α, were used to obtain the C_{eff} using Equation (3). The resistance (R) and capacitance (C) values were normalized to cell geometry dimensions and calculated using Equations (4) and (5), respectively. The conductivity (σ) and relative permittivity (ε_r) were also determined using Equations (6) and (7), respectively. The values were obtained using the equations below, where A_{Au} is the area of the gold electrodes, d_{Sample} is the thickness of the analyzed OIH film sample, and ε_0 stands for the vacuum permittivity in nF cm^{-1}:

$$R = R_{sample} \times A_{Au\ disc} \tag{4}$$

$$C = \frac{C_{eff}}{A_{Au\ disc}} \tag{5}$$

$$\sigma = \frac{d_{sample}}{A_{Au\ disc}} / R_{sample} \tag{6}$$

$$\varepsilon_r = \frac{C_{eff} \times d_{sample}}{\varepsilon_0} \times A_{Au\ disc} \tag{7}$$

The information regarding the electrical properties of the OIH films produced, including normalized resistance (R), capacitance (C), conductivity (σ), and relative permittivity (ε_r), is displayed in Table 2.

Table 2. Electrical properties of the OIH samples based on A(X) matrices. For comparison purposes the data reported for A(400) [35], A(600) [35], A(900) [35], and A(2000) [35] were also included.

OIH Sample	Molecular Weight	log R (Ω cm^2)	ε_r	C (nF cm^{-2})	$-$log σ (S cm^{-1})
A(270)	270	10.05 ± 0.02	17.80 ± 1.53	0.008 ± 0.001	10.76 ± 0.02
A(400) [35]	430	8.69 ± 0.10 [35]	26.90 ± 2.36 [35]	0.035 ± 0.002 [35]	9.85 ± 0.06 [35]
A(403)	440	8.48 ± 0.02	12.56 ± 0.12	$0.0086 \pm 8.37 \times 10^{-5}$	9.37 ± 0.02
A(600) [35]	600	5.97 ± 0.34 [35]	n.a. [35]	n.a. [35]	6.39 ± 0.90 [35]
A(900) [35]	900	4.78 ± 0.13 [35]	28.50 ± 4.00 [35]	0.041 ± 0.006 [35]	5.99 ± 0.05 [35]
A(170)	1700	7.03 ± 0.72	8.12 ± 0.03	$0.005 \pm 2.10 \times 10^{-5}$	7.87 ± 0.72
A(2000) [35]	2000	4.78 ± 0.29 [35]	56.75 ± 9.39 [35]	0.051 ± 0.009 [35]	5.89 ± 0.40 [35]
A(2001)	2050	9.76 ± 0.12	24.01 ± 2.26	0.013 ± 0.001	10.55 ± 0.12

n.a.—not applied.

Table 2 shows that for the OIHs A(270), A(400), A(403), A(600), A(900), and A(2000), as the molecular weight of the Jeffamines® used in the synthesis increases, the normalized resistance (R) decreases, which is in accordance with the literature [22,35]. The difference obtained for the resistance of A(270) compared to A(400) may be due to the fact that Jeffamine® RFD-270 is an aliphatic amine containing both rigid (cycloaliphatic) and flexible (etheramine) segments in the same molecule (vide Figure 1). In the case of A(400), the Jeffamine® used in the synthesis (Jeffamine® D-400) is a difunctional primary amine. This polyetheramine is characterized by repeating oxypropylene units in the backbone. This difference between the backbones of the Jeffamines® used (i.e., RFD-270 and D-400) may explain the two orders of magnitude difference found between the resistance of A(270) and A(400) and the capacitance that in case of A(400) is 4.375 times higher than A(270). Regarding the A(170) samples, a lower resistance was expected considering the molecular weight of the Jeffamine® used (1700 g mol^{-1}). Previous publications showed that as the molecular weight of the Jeffamine® increases, the resistance of the OIHs decreases due to the organic component increase [22,35]. Jeffamine® THF-170 is chemically based on a [poly(tetramethylene ether glycol)]/(poly(propylene glycol)) copolymer that contains a significant amount of secondary as well as primary amines. Therefore, this might be the main reason for the high resistance obtained. The main difference between the resistance obtained for A(2000) and A(2001) can be explained by the disparity between the Jeffamine® used in each synthesis. A(2001) was synthesized using a difunctional secondary amine (Jeffamine® SD-2001) while A(2000) was obtained using a polyether diamine based predominantly on a polyethylene glycol (PEG) backbone. Jeffamine® SD-2001 is a difunctional secondary amine derived from the Jeffamine® D-2000 amine, which are polyether diamines based on a poly(propylene glycol) (PPG) backbone. This may explain the main differences found between A(2000) and A(2001) regarding the electrical properties (Table 2) in which the resistance of A(2001) increases 2.04 times compared to the resistance obtained for A(2000).

The new OIH materials synthesized show a ε_r between 8 and 24, with A(170) and A(2001) showing the lowest and the highest values, respectively. The capacitance values obtained are between 0.008 and 0.013 nF cm^{-2}, with A(170) and A(2001) showing the lowest and the highest, respectively. Moreover, the new amino-alcohol-based sol-gel materials show lower ε_r and capacitance values than the ones already reported.

3.4. Preliminary Assessment of the OIH in Contact with Cement Paste

A preliminary assessment of the OIH materials in contact with a cement paste prepared with a w/c = 0.5 was conducted. The preliminary tests were only performed for the OIHs with the highest transmittance values, i.e., A(403), A(270), and A(170). The main objective of these studies was to assess if these new OIHs were stable and resistant when in contact with a high alkaline environment. A cement paste with a w/c = 0.5 was chosen to mimic the concrete alkalinity. The samples were immersed in the cement paste for 7, 14, and for 28 days. After 7, 14, and 28 days the cured paste samples were broken to check the physical aspect of the samples. Figure 7 shows the samples A(270), A(403), and A(170) after 7 days of being embedded in the mentioned cement paste.

Figure 7. OIH samples (A(270), A(403), and A(170)) after 7 days immersed in cement paste with a w/c = 0.5.

Figure 8 shows the results obtained for the OIH samples after 28 days of being embedded in the cement paste.

After 7 days of contact with the cement paste (Figure 7), except for OIH sample A(170), OIH A(270) and A(403) suffered severe degradation and reacted with the cement paste. Even though sample A(403) seems to be less resistant than A(270), due to the alkalinity of the cement pastes, it was not possible to extract OIH samples A(270) and A(403) from the cement paste after 7 days, as can be observed in Figure 7. Sample A(170) appears to be resistant to the high alkaline environment of the cement paste and did not experience visible degradation. It can be observed that after 28 days the same behavior was found. OIH sample A(170) showed improved stability when compared to the other samples. The promising preliminary results suggest that further tests such as EIS should be conducted. By performing EIS measurements, i.e., assessing resistance and capacitance of the A(170) samples before and after immersion in the cement pastes, it will be possible to quantify the degradation impact of the alkaline environment.

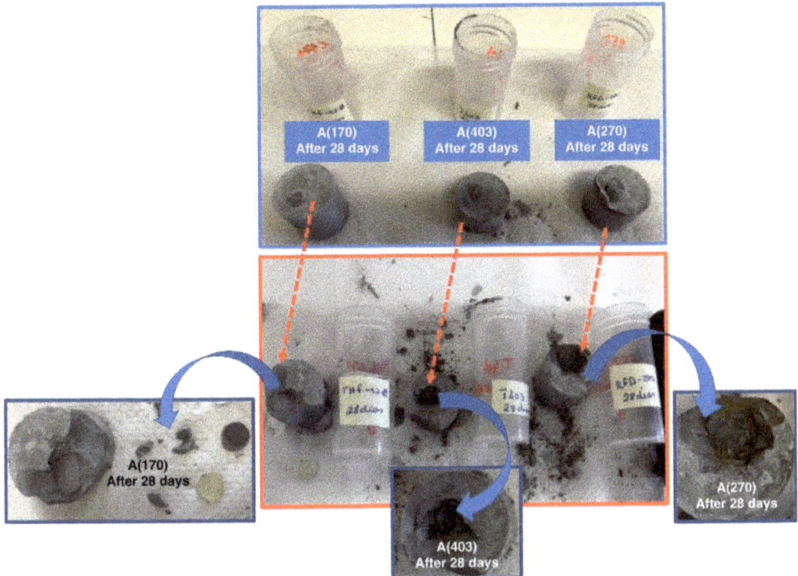

Figure 8. OIH samples (A(270), A(403), and A(170)) after 28 days immersed in cement paste with a w/c = 0.5.

Considering the results obtained, the thickness of the OIH A(170) films was measured before and after 7, 14, and 28 days of being embedded in the cement paste. Table 3 shows the thicknesses obtained for each A(170) OIH film tested.

Table 3. Thickness of the OIH films measured after 7, 14, and 28 days embedded in cement paste with a w/c = 0.5.

OIH Sample	Thickness/mm			
	Before Exposure to Cement Paste	After 7 Days in Cement Paste	After 14 Days in Cement Paste	After 28 Days in Cement Paste
A(170)	1.387	1.343 ($\Delta = 0.044$)		
	1.491		1.472 ($\Delta = 0.019$)	
	1.802			1.798 ($\Delta = 0.004$)

Table 3 shows that thickness decreases slightly over time for the A(170) samples. After 28 days a decrease of 0.22% of thickness was found. The highest decrease percentage was found for the OIH sample exposed for 7 days (around 3.17%), followed by the sample exposed for 14 days, which lost around 1.27% thickness.

4. Conclusions

The synthesis of the five new OIH sol-gel materials, homogeneous and crack-free, based on amino-alcohol silicate matrices, namely A(148), A(270), A(403), A(170), and A(2001), has been reported. The results obtained allow for the conclusion that the different Jeffamines® used influence the optical and the electrochemical properties. The OIH matrices with the lowest and highest transmittance, between 400 and 700 nm, were given by A(2001) and A(403), respectively. The photoluminescence spectra obtained for xerogel films A(148), A(270), A(403), and A(170) allow for the conclusion that these materials show an intrinsic emission and provide generally shorter polyether chains than samples with higher photoluminescence. The capacitance values obtained were lower than the ones already reported for similar materials and ranged between 0.008 and 0.013 nF cm^{-2}. The A(170) samples showed good behavior in contact with cement paste, allowing for the conclusion that only A(170) samples are resistant to highly alkaline environments. Therefore, considering the optical and electrical properties and the chemical stability of A(170) samples in contact with cement paste, it can be concluded that these samples show promising properties as support films in optical sensor fields such as fiber sensor devices for application in the civil engineering field. Regarding the A(270) and A(403) samples, it can be concluded that, despite the interesting optical and electrical properties, these OIH samples are not suitable to be used in highly alkaline environments. Moreover, further studies should be conducted on the A(170) samples in order to quantify the degradation impact due to the alkaline environment.

Author Contributions: Conceptualization, R.B.F. and C.J.R.S.; formal analysis, R.B.F, B.R.G., S.P.G.C., C.J.R.S., and M.M.M.R; funding acquisition, R.B.F. and C.J.R.S.; investigation, R.B.F. and B.R.G.; methodology, R.B.F. and C.J.R.S.; writing—original draft preparation, R.B.F.; writing—review and editing, R.B.F, B.R.G., S.P.G.C., C.J.R.S., and M.M.M.R.; project administration, R.B.F. and C.J.R.S.; resources, C.J.R.S. and R.B.F.; supervision, R.B.F. All authors have read and agreed to the published version of the manuscript with the exception of C.J.R.S.

Funding: This research was funded by the "SolSensors—Development of Advanced Fiber Optic Sensors for Monitoring the Durability of Concrete Structures" project, with the Program Budget COMPETE—Operational Program Competitiveness and Internationalization—COMPETE 2020 reference and the Lisbon Regional Operational Program (its FEDER component).

Acknowledgments: The authors acknowledge the reviewers for the constructive comments and suggestions and the support of Centro de Química, CQUM, which is financed by national funds through the FCT Foundation for Science and Technology, I.P. under project UID/QUI/00686/2020.

Conflicts of Interest: The authors declare no conflict of interest. The funders had no role in the design of the study; in the collection, analyses, or interpretation of data; in the writing of the manuscript, or in the decision to publish the results.

References

1. Bekiari, V.; Lianos, P. Tunable Photoluminescence from a Material Made by the Interaction between (3-Aminopropyl)triethoxysilane and Organic Acids. *Chem. Mater.* **1998**, *10*, 3777–3779. [CrossRef]

2. Carlos, L.D.; de Zea Bermudez, V.; Sá Ferreira, R.A.; Marques, L.; Assunção, M. Sol–Gel Derived Urea Cross-Linked Organically Modified Silicates. 2. Blue-Light Emission. *Chem. Mater.* **1999**, *11*, 581–588. [CrossRef]

3. Bekiari, V.; Lianos, P. Multicolor emission from terpyridine–lanthanide ion complexes encapsulated in nanocomposite silica/poly(ethylene glycol) sol–gel matrices. *J. Lumin.* **2003**, *101*, 135–140. [CrossRef]

4. Kober, U.A.; Gallas, M.R.; Campo, L.F.; Rodembusch, F.S.; Stefani, V. Fluorescence emission modulation in singlefluoroforic submicro-sized silica particles. *J. Sol-Gel Sci. Technol.* **2009**, *52*, 305–308. [CrossRef]

5. Pletsch, D.; da Santos, F.S.; Severo Rodembusch, F.; Stefani, V.; Franciscato Campo, L. Bis-silylated terephthalate as a building block precursor for highly fluorescent organic–inorganic hybrid materials. *New J. Chem.* **2012**, *36*, 2506–2513. [CrossRef]

6. Ribeiro, S.J.L.; Dahmouche, K.; Ribeiro, C.A.; Santilli, C.V.; Pulcinelli, S.H. Study of Hybrid Silica-Polyethyleneglycol Xerogels by Eu^{3+} Luminescence Spectroscopy. *J. Sol-Gel Sci. Technol.* **1998**, *13*, 427–432. [CrossRef]

7. Bekiari, V.; Lianos, P.; Judeinstein, P. Efficient luminescent materials made by incorporation of terbium(III) and 2,2-bipyridine in silica/poly(ethylene oxide) hybrid gels. *Chem. Phys. Lett.* **1999**, *307*, 310–316. [CrossRef]

8. Severo Rodembusch, F.; Franciscato Campo, L.; Stefani, V.; Rigacci, A. The first silica aerogels fluorescent by excited state intramolecular proton transfer mechanism (ESIPT). *J. Mater. Chem.* **2005**, *15*, 1537–1541. [CrossRef]

9. Grando, S.R.; Pessoa, C.M.; Gallas, M.R.; Costa, T.M.H.; Rodembusch, F.S.; Benvenutti, E.V. Modulation of the ESIPT Emission of Benzothiazole Type Dye Incorporated in Silica-Based Hybrid Materials. *Langmuir* **2009**, *25*, 13219–13223. [CrossRef]

10. Barja, B.C.; Bari, S.E.; Marchi, M.C.; Iglesias, F.L.; Bernardi, M. Luminescent Eu(III) hybrid sensors for in situ copper detection. *Sens. Actuators B Chem.* **2011**, *158*, 214–222. [CrossRef]

11. Nolasco, M.M.; Vaz, P.M.; Freitas, V.T.; Lima, P.P.; André, P.S.; Ferreira, R.A.S.; Vaz, P.D.; Ribeiro-Claro, P.; Carlos, L.D. Engineering highly efficient Eu(III)-based tri-ureasil hybrids toward luminescent solar concentrators. *J. Mater. Chem. A* **2013**, *1*, 7339–7350. [CrossRef]

12. Brito, J.B.; Costa, T.M.H.; Rodembusch, F.S.; Konowalow, A.S.; dos Reis, R.M.S.; Balzaretti, N.M.; da Jornada, J.A.H. Blue–green luminescent carbon nanodots produced in a silica matrix. *Carbon* **2015**, *91*, 234–240. [CrossRef]

13. He, W.; Xie, Y.; Xing, Q.; Ni, P.; Han, Y.; Dai, H. Sol-gel synthesis of biocompatible Eu^{3+}/Gd^{3+} co-doped calcium phosphate nanocrystals for cell bioimaging. *J. Lumin.* **2017**, *192*, 902–909. [CrossRef]

14. de Zea Bermudez, V.; Carlos, L.D.; Alcácer, L. Sol–Gel Derived Urea Cross-Linked Organically Modified Silicates. 1. Room Temperature Mid-Infrared Spectra. *Chem. Mater.* **1999**, *11*, 569–580. [CrossRef]

15. Boev, V.I.; Soloviev, A.; Silva, C.J.R.; Gomes, M.J.M.; Barber, D.J. Highly transparent sol-gel derived ureasilicate monoliths exhibiting long-term optical stability. *J. Sol-Gel Sci. Technol.* **2006**, *41*, 223–229. [CrossRef]

16. Boev, V.I.; Soloviev, A.; Silva, C.J.R.; Gomes, M.J.M. Incorporation of CdS nanoparticles from colloidal solution into optically clear ureasilicate matrix with preservation of quantum size effect. *Solid State Sci.* **2006**, *8*, 50–58. [CrossRef]

17. Gonçalves, L.F.F.F.; Silva, C.J.R.; Kanodarwala, F.K.; Stride, J.A.; Pereira, M.R.; Gomes, M.J.M. Synthesis and characterization of organic–inorganic hybrid materials prepared by sol–gel and containing ZnxCd1−xS nanoparticles prepared by a colloidal method. *J. Lumin.* **2013**, *144*, 203–211. [CrossRef]

18. Moreira, S.D.F.C.; Silva, C.J.R.; Prado, L.A.S.A.; Costa, M.F.M.; Boev, V.I.; Martín-Sánchez, J.; Gomes, M.J.M. Development of new high transparent hybrid organic-inorganic monoliths with surface engraved diffraction pattern. *J. Polym. Sci. Part B Polym. Phys.* **2012**, *50*, 492–499. [CrossRef]

19. Figueira, R.B.; Silva, C.J.; Pereira, E.V.; Salta, M.M. Ureasilicate Hybrid Coatings for Corrosion Protection of Galvanized Steel in Cementitious Media. *J. Electrochem. Soc.* **2013**, *160*, C467–C479. [CrossRef]

20. Gonçalves, L.F.F.F.; Silva, C.J.R.; Kanodarwala, F.K.; Stride, J.A.; Pereira, M.R.; Gomes, M.J.M. Synthesis of an optically clear, flexible and stable hybrid ureasilicate matrix doped with CdSe nanoparticles produced by reverse micelles. *Mater. Chem. Phys.* **2014**, *147*, 86–94. [CrossRef]

21. Molina, E.F.; Parreira, R.L.T.; De Faria, E.H.; de Carvalho, H.W.P.; Caramori, G.F.; Coimbra, D.F.; Nassar, E.J.; Ciuffi, K.J. Ureasil-Poly(ethylene oxide) Hybrid Matrix for Selective Adsorption and Separation of Dyes from Water. *Langmuir* **2014**, *30*, 3857–3868. [CrossRef] [PubMed]

22. Figueira, R.B.; Silva, C.J.R.; Pereira, E.V. Hybrid sol–gel coatings for corrosion protection of hot-dip galvanized steel in alkaline medium. *Surf. Coat. Technol.* **2015**, *265*, 191–204. [CrossRef]

23. Correia, S.F.H.; Antunes, P.; Pecoraro, E.; Lima, P.P.; Varum, H.; Carlos, L.D.; Ferreira, R.A.S.; André, P.S. Optical Fiber Relative Humidity Sensor Based on a FBG with a Di-Ureasil Coating. *Sensors* **2012**, *12*, 8847–8860. [CrossRef] [PubMed]

24. de Jesus, N.A.M.; de Oliveira, A.H.P.; Tavares, D.C.; Furtado, R.A.; de Silva, M.L.A.; Cunha, W.R.; Molina, E.F. Biofilm formed from a tri-ureasil organic–inorganic hybrid gel for use as a cubebin release system. *J. Sol-Gel Sci. Technol.* **2018**, *88*, 192–201. [CrossRef]

25. Nunes, S.C.; Fernandes, M.; Gonçalves, H.M.R.; Serrano, J.L.; Almeida, P.; de Bermudez, V.Z. Di-urea cross-linked siloxane hybrid materials incorporating oligo(oxypropylene) and oligo(oxyethylene) chains. *J. Sol-Gel Sci. Technol.* **2020**, 1–15. [CrossRef]

26. Gonçalves, L.F.F.F.; Kanodarwala, F.K.; Stride, J.A.; Silva, C.J.R.; Gomes, M.J.M. One-pot synthesis of CdS nanoparticles exhibiting quantum size effect prepared within a sol–gel derived ureasilicate matrix. *Opt. Mater.* **2013**, *36*, 186–190. [CrossRef]

27. Gonçalves, L.F.F.F.; Silva, C.J.R.; Kanodarwala, F.K.; Stride, J.A.; Gomes, M.J.M. Synthesis and characterization of organic–inorganic hybrid materials prepared by sol–gel and containing CdS nanoparticles prepared by a colloidal method using poly(N-vinyl-2-pyrrolidone). *J. Sol-Gel Sci. Technol.* **2014**, *71*, 69–78. [CrossRef]

28. Gonçalves, L.F.F.F.; Silva, C.J.R.; Kanodarwala, F.K.; Stride, J.A.; Pereira, M.R.; Gomes, M.J.M. Influence of Cd^{2+}/S^{2-} molar ratio and of different capping environments in the optical properties of CdS nanoparticles incorporated within a hybrid diureasil matrix. *Appl. Surf. Sci.* **2014**, *314*, 877–887. [CrossRef]

29. Gonçalves, L.F.F.F.; Kanodarwala, F.K.; Stride, J.A.; Silva, C.J.R.; Pereira, M.R.; Gomes, M.J.M. One-pot synthesis of CdSe nanoparticles exhibiting quantum size effect within a sol–gel derived ureasilicate matrix. *J. Photochem. Photobiol. Chem.* **2014**, *285*, 21–29. [CrossRef]

30. Gonçalves, J.T.; Boev, V.I.; Solovyev, A.; Silva, C.J.R.; Gomes, M.J.M. Luminescence and Absorption of Hybrid Xerogels Doped with PbS Nanoparticles Prepared by Gas Diffusion Method. *Mater. Sci. Forum* **2006**, *514–516*, 1221–1224. [CrossRef]

31. Figueira, R.B.; Silva, C.J.R.; Pereira, E.V. Influence of Experimental Parameters Using the Dip-Coating Method on the Barrier Performance of Hybrid Sol-Gel Coatings in Strong Alkaline Environments. *Coatings* **2015**, *5*, 124–141. [CrossRef]

32. Figueira, R.B.; Silva, C.J.; Pereira, E.V. Ureasilicate Hybrid Coatings for Corrosion Protection of Galvanized Steel in Chloride-Contaminated Simulated Concrete Pore Solution. *J. Electrochem. Soc.* **2015**, *162*, C666–C676. [CrossRef]

33. Figueira, R.B.; Callone, E.; Silva, C.J.R.; Pereira, E.V.; Dirè, S. Hybrid Coatings Enriched with Tetraethoxysilane for Corrosion Mitigation of Hot-Dip Galvanized Steel in Chloride Contaminated Simulated Concrete Pore Solutions. *Materials* **2017**, *10*, 306. [CrossRef] [PubMed]

34. Willis-Fox, N.; Kraft, M.; Arlt, J.; Scherf, U.; Evans, R.C. Tunable White-Light Emission from Conjugated Polymer-Di-Ureasil Materials. *Adv. Funct. Mater.* **2016**, *26*, 532–542. [CrossRef]

35. Figueira, R.B.; Silva, C.J.; Pereira, E.V.; Salta, M.M. Alcohol-Aminosilicate Hybrid Coatings for Corrosion Protection of Galvanized Steel in Mortar. *J. Electrochem. Soc.* **2014**, *161*, C349–C362. [CrossRef]

36. Cho, N.; Kim, N.; Jang, J.; Chang, S. Estimation of deflection curve of bridges using fiber optic strain sensors. In Proceedings of the Smart Structures and Materials 2000: Smart Systems for Bridges, Structures, and Highways; International Society for Optics and Photonics, Newport Beach, CA, USA, 6 March 2000; SPIE: Bellingham, WA, USA, 2000; Volume 3988, pp. 339–348.

37. Kim, N.-S.; Cho, N.-S. Estimating deflection of a simple beam model using fiber optic bragg-grating sensors. *Exp. Mech.* **2004**, *44*, 433–439. [CrossRef]

38. Yehia, S.; Landolsi, T.; Hassan, M.; Hallal, M. Monitoring of strain induced by heat of hydration, cyclic and dynamic loads in concrete structures using fiber-optics sensors. *Measurement* **2014**, *52*, 33–46. [CrossRef]

39. Marković, M.Z.; Bajić, J.S.; Batilović, M.; Sušić, Z.; Joža, A.; Stojanović, G.M. Comparative Analysis of Deformation Determination by Applying Fiber-optic 2D Deflection Sensors and Geodetic Measurements. *Sensors* **2019**, *19*, 844. [CrossRef]

40. Wan, K.T.; Leung, C.K.Y. Applications of a distributed fiber optic crack sensor for concrete structures. *Sens. Actuators Phys.* **2007**, *135*, 458–464. [CrossRef]

41. Bao, T.; Wang, J.; Yao, Y. A fiber optic sensor for detecting and monitoring cracks in concrete structures. *Sci. China Technol. Sci.* **2010**, *53*, 3045–3050. [CrossRef]

42. Luo, D.; Yue, Y.; Li, P.; Ma, J.; Zhang, L.L.; Ibrahim, Z.; Ismail, Z. Concrete beam crack detection using tapered polymer optical fiber sensors. *Measurement* **2016**, *88*, 96–103. [CrossRef]

43. Baccay, M.A.; Otsuki, N.; Nishida, T.; Maruyama, S. Influence of Cement Type and Temperature on the Rate of Corrosion of Steel in Concrete Exposed to Carbonation. *Corrosion* **2006**, *62*, 811–821. [CrossRef]

44. Blanc, P.h.; Bourbon, X.; Lassin, A.; Gaucher, E.C. Chemical model for cement-based materials: Temperature dependence of thermodynamic functions for nanocrystalline and crystalline C–S–H phases. *Cem. Concr. Res.* **2010**, *40*, 851–866. [CrossRef]

45. Behnood, A.; Van Tittelboom, K.; De Belie, N. Methods for measuring pH in concrete: A review. *Constr. Build. Mater.* **2016**, *105*, 176–188. [CrossRef]

46. Plusquellec, G.; Geiker, M.R.; Lindgård, J.; Duchesne, J.; Fournier, B.; De Weerdt, K. Determination of the pH and the free alkali metal content in the pore solution of concrete: Review and experimental comparison. *Cem. Concr. Res.* **2017**, *96*, 13–26. [CrossRef]

47. Wei, Y.; Guo, W.; Zheng, X. Integrated shrinkage, relative humidity, strength development, and cracking potential of internally cured concrete exposed to different drying conditions. *Dry. Technol.* **2016**, *34*, 741–752. [CrossRef]

48. Multon, S.; Toutlemonde, F. Effect of moisture conditions and transfers on alkali silica reaction damaged structures. *Cem. Concr. Res.* **2010**, *40*, 924–934. [CrossRef]

49. Zhang, W.; Min, H.; Gu, X. Temperature response and moisture transport in damaged concrete under an atmospheric environment. *Constr. Build. Mater.* **2016**, *123*, 290–299. [CrossRef]

50. Figueira, R.B.; Sadovski, A.; Melo, A.P.; Pereira, E.V. Chloride threshold value to initiate reinforcement corrosion in simulated concrete pore solutions: The influence of surface finishing and pH. *Constr. Build. Mater.* **2017**, *141*, 183–200. [CrossRef]

51. Alonso, C.; Andrade, C.; Castellote, M.; Castro, P. Chloride threshold values to depassivate reinforcing bars embedded in a standardized OPC mortar. *Cem. Concr. Res.* **2000**, *30*, 1047–1055. [CrossRef]

52. Angst, U.; Elsener, B.; Larsen, C.K.; Vennesland, Ø. Critical chloride content in reinforced concrete—A review. *Cem. Concr. Res.* **2009**, *39*, 1122–1138. [CrossRef]

53. Figueira, R.B.; Sousa, R.; Coelho, L.; Azenha, M.; de Almeida, J.M.; Jorge, P.A.S.; Silva, C.J.R. Alkali-silica reaction in concrete: Mechanisms, mitigation and test methods. *Constr. Build. Mater.* **2019**, *222*, 903–931. [CrossRef]

54. Figueira, R.B. Electrochemical Sensors for Monitoring the Corrosion Conditions of Reinforced Concrete Structures: A Review. *Appl. Sci.* **2017**, *7*, 1157. [CrossRef]

55. Spírková, M.; Brus, J.; Hlavatá, D.; Kamisová, H.; Matejka, L.; Strachota, A. Preparation and characterisation of hybrid organic/inorganic coatings and films. *Surf. Coat. Int. Part B Coat. Trans.* **2003**, *86*, 187–193. [CrossRef]

56. Špírková, M.; Brus, J.; Hlavatá, D.; Kamišová, H.; Matějka, L.; Strachota, A. Preparation and characterization of hybrid organic–inorganic epoxide-based films and coatings prepared by the sol–gel process. *J. Appl. Polym. Sci.* **2004**, *92*, 937–950. [CrossRef]

57. Simcha, S.; Dotan, A.; Kenig, S.; Dodiuk, H. Characterization of Hybrid Epoxy Nanocomposites. *Nanomaterials* **2012**, *2*, 348–365. [CrossRef]

58. Alhwaige, A.A.; Alhassan, S.M.; Katsiotis, M.S.; Ishida, H.; Qutubuddin, S. Interactions, morphology and thermal stability of graphene-oxide reinforced polymer aerogels derived from star-like telechelic aldehyde-terminal benzoxazine resin. *RSC Adv.* **2015**, *5*, 92719–92731. [CrossRef]

59. Meazzini, I.; Willis-Fox, N.; Blayo, C.; Arlt, J.; Clément, S.; Evans, R.C. Targeted design leads to tunable photoluminescence from perylene dicarboxdiimide–poly(oxyalkylene)/siloxane hybrids for luminescent solar concentrators. *J. Mater. Chem. C* **2016**, *4*, 4049–4059. [CrossRef]

60. Erdem, A.; Ngwabebhoh, F.A.; Yildiz, U. Synthesis, characterization and swelling investigations of novel polyetheramine-based hydrogels. *Polym. Bull.* **2017**, *74*, 873–893. [CrossRef]

61. Carlos, L.D.; Sá Ferreira, R.A.; Pereira, R.N.; Assunção, M.; de Zea Bermudez, V. White-Light Emission of Amine-Functionalized Organic/Inorganic Hybrids: Emitting Centers and Recombination Mechanisms. *J. Phys. Chem. B* **2004**, *108*, 14924–14932. [CrossRef]

62. Bekiari, V.; Lianos, P.; Stangar, U.L.; Orel, B.; Judeinstein, P. Optimization of the Intensity of Luminescence Emission from Silica/Poly(ethylene oxide) and Silica/Poly(propylene oxide) Nanocomposite Gels. *Chem. Mater.* **2000**, *12*, 3095–3099. [CrossRef]

63. Cordoncillo, E.; Guaita, F.J.; Escribano, P.; Philippe, C.; Viana, B.; Sanchez, C. Blue emitting hybrid organic–inorganic materials. *Opt. Mater.* **2001**, *18*, 309–320. [CrossRef]

64. Santos, T.C.F.; Rodrigues, R.R.; Correia, S.F.H.; Carlos, L.D.; Ferreira, R.A.S.; Molina, C.; Péres, L.O. UV-converting blue-emitting polyfluorene-based organic-inorganic hybrids for solid state lighting. *Polymer* **2019**, *174*, 109–113. [CrossRef]

65. Yasakau, K.A.; Ferreira, M.G.S.; Zheludkevich, M.L. Sol-Gel Coatings with Nanocontainers of Corrosion Inhibitors for Active Corrosion Protection of Metallic Materials. In *Handbook of Sol-Gel Science and Technology*; Klein, L., Aparicio, M., Jitianu, A., Eds.; Springer International Publishing: Berlin, Germany, 2017; pp. 1–37, ISBN 978-3-319-19454-7.

66. Orazem, M.E.; Tribollet, B. *Electrochemical Impedance Spectroscopy*; Wiley: Hoboken, NJ, USA, 2008; ISBN 978-0-470-38157-1.

67. Barsoukov, E.; Macdonald, J.R. Impedance spectroscopy: Theory, experiment, and applications. *History* **2005**, *1*, 1–13.

68. Zoltowski, P. On the electrical capacitance of interfaces exhibiting constant phase element behaviour. *J. Electroanal. Chem.* **1998**, *443*, 149–154. [CrossRef]

Publisher's Note: MDPI stays neutral with regard to jurisdictional claims in published maps and institutional affiliations.

 *polymers*

Article

Preparation and Properties of Fluorosilicone Fouling-Release Coatings

Tong Wu, Yuhong Qi *, Qi'an Chen , Chuanjun Gu and Zhanping Zhang

Department of Materials Science and Engineering, Dalian Maritime University, Dalian 116026, China
* Correspondence: yuhong_qi@dlmu.edu.cn

Abstract: To improve the antifouling performance of silicone fouling-release coatings, some fluorosilicone and silicone fouling-release coatings were prepared and cured at room temperature with hydroxyl-terminated fluoropolysiloxane (FPS) or hydroxy-terminated polydimethylsiloxane (PDMS) as a film-forming resin, tetraethyl orthosilicate (TEOS) as a crosslinking agent, and dibutyltin dilaurate (DBTDL) as a catalyst. The chemical structure, surface morphology and roughness, tensile properties, and antifouling properties of the coating were studied by infrared spectroscopy, a laser confocal scanning microscope, contact angle measurement, tensile tests, and marine bacteria and benthic diatom attachment tests. The results showed that the FPS coatings were not only hydrophobic but also oleophobic, and the contact angles of the FPS coatings were larger than those of the PDMS coatings. The surface free energies of the FPS coatings were much lower than those of the PDMS coatings. Generally, the fluorine groups can improve the antifouling performance of the coating. Introducing nonreactive silicone oil into PDMS or FPS coatings can improve the antifouling performance of the coating to a certain extent. The prepared fluorosilicone fouling-release coatings showed good application prospects.

Keywords: fluorosilicone; hydroxyl terminated fluoropolysiloxane; antifouling; silicone oil; coatings

Citation: Wu, T.; Qi, Y.; Chen, Q.; Gu, C.; Zhang, Z. Preparation and Properties of Fluorosilicone Fouling-Release Coatings. *Polymers* **2022**, *14*, 3804. https://doi.org/10.3390/polym14183804

Academic Editors: Jesús-María García-Martínez and Emilia P. Collar

Received: 27 July 2022
Accepted: 8 September 2022
Published: 11 September 2022

Publisher's Note: MDPI stays neutral with regard to jurisdictional claims in published maps and institutional affiliations.

1. Introduction

Marine organisms such as microorganisms, barnacles, and algae adhere to the surface of ships and accumulate, resulting in an irregular distribution [1] and marine biological pollution. If the marine biological pollution is not controlled, the ships' surface roughness and resistance will increase, which will lead to a significant increase in fuel consumption, thereby increasing greenhouse gas emissions [2] and indirectly increasing the rate of global warming [3]. Simultaneously, marine organisms are introduced into nonnative environments through ship transportation [4], resulting in the transfer of invasive species, which not only causes damage to the ecological environment [5] but also affects economic development and navigation safety. Therefore, marine biological pollution has always been a problem for essential issues on a global scale [6,7]. To solve this problem, at first, coatings containing biocides are applied to the surface of ships. Still, such antifouling coatings use toxic and harmful materials such as amines, amides, organotin, and cuprous oxide, which have a malignant impact on the environment and accumulate in marine organisms [8]. The high use of toxic compounds has raised concerns [9,10], and such coatings have been gradually banned worldwide [3]. Various ecologically friendly antifouling coatings have been developed, such as amphiphilic antifouling coatings, protein-resistant polymers, antifouling release coatings, conductive antifouling coatings, and biomimetic antifouling coatings [11–15]. Among them, fouling-release coatings with low surface energy, high elasticity, and hydrophobicity have always been the focus of researchers. Typical FR coatings consist of fluoropolymers or polydimethylsiloxane (PDMS) [16], which has a soft siloxane backbone, good water stability [17], and low modulus [18]. Still, the oil resistance is unideal, it was found by research [19] that siloxane resins are dominated by CF_2 groups, and their stain resistance is much lower than that of CF_3 groups. Combining the high- and low-temperature resistance of silicone and the oil resistance of organic fluorine, the fluorine group acts as a hydrophobic group [20,21]. From the perspective of antifouling, the antifouling performance of the coating can be better.

The researchers' exploration of fluorosilicone resins has mainly focused on the introduction of fluorine groups into the polysiloxane network. The patent filed by Mera [22] produced a self-curing fluorosilicone resin crosslinked with a nonfluorinated polysiloxane resin, and the surface energy was lower than that of the pure silicone coating. Stafslien et al. [23] grafted a trifluoropropyl group (CF_3-PDMS) and polyethylene glycol pendant groups (TMS-PEG) on the network structure of polysiloxane at the concentrations of TMS-PEG and CF_3-PDMS as variables, and incorporating these two compounds into coatings was found to enhance the surface properties of coatings.

In this paper, hydroxyl-terminated fluoropolysiloxane was selected as the primary film-forming material to prepare the new fluorosilicone coating with three components. The main chain structure of the resin is the same as that of polydimethylsiloxane, which retains the excellent properties of the main chain of siloxane. Simultaneously, CF_3 groups were used to enhance the antifouling performance. Nonreactive silicone oil was used to improve the fouling-release performance, and inorganic powder was used to enhance the mechanical properties of the coating.

2. Materials and Methods

2.1. Materials

Hydroxyl-terminated fluoropolysiloxane (FPS, $HO[CH_3SiCH_2CH_2CF_3O]_nOH$) and methyl fluorosilicone oil (MFO, $Si(CH_3)_3O[CH_3SiCH_2CH_2CF_3O]_nSi(CH_3)_3$) were obtained from Shanghai Silicon Mountain Macromolecular Materials Co., Ltd. (Shanghai, China). The Brookfield viscosity of FPS is 10,000 cp. The Brookfield viscosity of MFO is 100 cp. Hydroxy-terminated polydimethylsiloxane (PDMS, $HOSi(CH_3)_2O[Si(CH_3)_2O]_nSi(CH_3)_2OH$) was obtained from Dayi Chemical Industry Co., Ltd. (Yantai, China). Its kinematic viscosity is 10,000 mm^2/s. Phenyl silicone oil (PSO, $Si(CH_3)_3O[Si(CH_3)_2O]_n[Si(C_6H_5)_2O]_mSi(CH_3)_3$) was obtained from the Damao Chemical Reagent Factory. Its kinematic viscosity is 30 mm^2/s. Tetraethyl orthosilicate (TEOS) was obtained from Shanghai Aladdin Biochemical Technology Co., Ltd. (Shanghai, China) as an analytical grade. Dibutyltin dilaurate (DBTDL) was obtained from Tianjin Kemiou Chemical Reagent Co., Ltd. (Tianjin, China) as an analytical grade.

2.2. Preparation of the Coating

The inorganic powder was dried in an oven at 120 °C for 48 h before the preparation of the paint. The coating consists of three components, namely Components A, B, and C. The formulation of Component A for prepared coatings is listed in Table 1. The preparation process of the studied coatings is shown in Figure 1. The crosslinking and curing mechanism of the fluorosilicone coatings is shown in Figure 2.

Table 1. Formulation of studied coatings (g).

	Component				A				B	C
	Material	FPS	PDMS	PSO	MFO	TiO_2	SiO_2	Xylene	TEOS	DBTDL
Coating	F-MFO	100			5	20		25	7.5	0.4
	F-PSO	100		5		20		25	7.5	0.4
	FPS	100				20		30	7.5	0.4
	P-MFO		100		5	10	10	25	15	0.4
	P-PSO		100	5		10	10	25	15	0.4
	PDMS		100			10	10	30	15	0.4

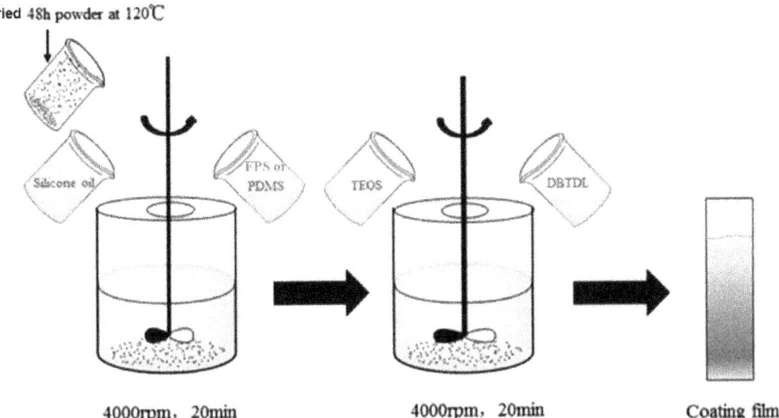

Figure 1. Schematic diagram of preparation process for studied coatings.

Figure 2. Crosslinking reaction mechanism of the fluorosilicone coatings.

First, the primary film-forming material FPS or PDMS and xylene were placed into the BGD750 grinding, dispersing, and stirring multi-purpose machine (Guangzhou Biuged Laboratory Instruments Co., Ltd., Guangzhou, China), and nonreactive silicone oil and powder was added in sequence according to the formula proportion and was mixed at 20 min. Then, the mixture was placed into the QZM conical mill (Tianjin Jingke Material Testing Machine Factory, Tianjin, China) and ground for 30 min, obtaining Component A. Then, TEOS and Component A were mixed, stirred evenly, and matured for 30 min; DBTDL was added into the mixture and stirred for 5 min. Finally, it was brushed on glass slides with a size of 75 mm × 25 mm × 1 mm and cured for subsequent evaluations, poured into a Teflon mold of 150 mm × 150 mm × 5 mm, and cured for 7 days in ambient for tensile tests. In order to eliminate the influence of coating thickness on the results, the same weight of each paint was used to prepare these samples.

2.3. Methods

2.3.1. Fourier Transform Infrared (FTIR) Spectroscopy

The molecular structure of the prepared coating was characterized by a Fourier Transfer Infrared Spectrometer (FTIR) (PerkinElmer Co., Ltd., Waltham, MA, USA) using the Attenuated Total Reflection (ATR) method. The scanning range was 4000–600 cm^{-1}. The resolution was 2 cm^{-1}. The number of scans was 32.

2.3.2. Morphology Analysis

The morphology of the coating surface was measured by an Olympus OLS4000 CLSM (OLYMPUS (China) Co., Ltd., Beijing, China), and the coating roughness (Sa) was measured by LEXT analysis software (version 2.2.4).

2.3.3. Contact Angle (CA) and Surface Energy

The water contact angle (WCA) and diiodomethane contact angle (DCA) on the studied coatings were measured using a JC2000C contact angle meter (Shanghai Zhongchen Digital Technic Apparatus Co., Ltd., Shanghai, China). The droplet size was 3 µL, and the image was taken immediately after the droplet was in contact with the coating for 3 s. The measurement calculation was performed with 6 droplets for each coating, and the average value was taken. According to the two-liquid method proposed by Owens [24], the surface free energy of each coating was calculated with Formulas (1)–(3). In the formulas, θ_{H_2O} is WCA; $\theta_{CH_2I_2}$ is DCA; σ_S^p is the polar force; σ_S^d is the dispersion force; σ_S is the surface free energy.

$$\sigma_s^p = [(137.5 + 256.1 \times \cos\theta_{H_2O}\ 118.6 \times \cos\theta_{CH_2I_2})/44.92]^2 \tag{1}$$

$$\sigma_s^d = [(139.9 + 181.4 \times \cos\theta_{CH_2I_2}\ 41.5 \times \cos\theta_{H_2O})/44.92]^2 \tag{2}$$

$$\sigma_s = \sigma_s^p + \sigma_s^d \tag{3}$$

2.3.4. Tensile Test

According to Chinese standard GB/T528-2009 (ISO37-2005) [25], the UTM5105 microcomputer-controlled electronic universal testing machine (Jinan Wance Electrical Equipment Co., Ltd., Jinan, China) was used to test the tensile properties of the samples. Three specimens were used for each coating, and the tensile speed was set to 50 mm/min. The data were analyzed to obtain the elastic modulus and stress at 100% elongation of the coating.

2.3.5. Antifouling Evaluation
Marine Bacterial Adhesion Test

The natural seawater obtained from the Yellow Sea of Dalian in China was used for the marine bacterial adhesion test. Before the test, all utensils were sterilized in the pressure steam for 20 min, and then they were placed on the SW-CJ-1FD clean workbench for 20 min of ultraviolet disinfection.

The test process was as follows. First, 6 specimens of each coating were completely immersed in fresh seawater and incubated under 28 °C for 12 h in light and 12 h in dark. After 24 h, the samples were taken out. All 6 samples were rinsed gently in sterilized seawater to release unattached bacteria on the surface. Then, 3 samples were taken out and placed into a 50 mL centrifuge tube containing 40 mL of sterilized seawater, and the HY-4 speed-regulated multi-purpose oscillator was used to simulate the surface state of the seawater washing the coating. The oscillator parameters were as follows: vibration amplitude of 20 mm, oscillation frequency of 130 rpm, and oscillation time of 15 min. These three treated samples were named as Washed. The other three untreated ones were named as Rinsed. Then, the bacterial film was brushed on the surface of the sample into sterilized seawater and diluted 1 million times. Then, 10 µL of the diluent was taken and spread evenly on the 2216E solid medium, its formula was listed in Table 2. Next, the medium was placed into the biochemical incubator (SHP-080) and cultivated at 25 °C. It was observed once every 24 h and the number of colonies on the culture medium was recorded at 24 h and 48 h. The bacteria colony on the medium at 48 h was photographed and reported as test results. The images were quantified by Image-Pro Plus software. The average and standard deviation of the number of colonies on the three media of each group of samples were taken, and the bacterial removal rate (R) was calculated with Formula (4).

$$R = \frac{(C_{Rinsed} - C_{Washed})}{C_{Rinsed}} \times 100\% \tag{4}$$

Table 2. Formula of 2216E solid medium.

Ingredients	Peptone	Yeast Extract	FePO$_4$	Agar	Sterilized Seawater
Content	2 g	0.4 g	0.004 g	8 g	400 mL

Navicula Tenera Adhesion Test

First, fresh seawater was used to prepare *Navicula Tenera* solution with an algal concentration of 10^{-5}–10^{-6} cell/mL, and 6 samples were placed in the solution and cultured at 22 °C under the control of the golden ratio of 12 h:12 h. After 24 h, the samples were taken to determine the chlorophyll a value of *Navicula Tenera* attached to the coating surface. The specific methods for the culture of *Navicula Tenera* can be found in the literature [26,27]. Six samples for each coating were also treated as mentioned above and divided into rinsed and washed. To extract chlorophyll a of *Navicula Tenera* attached to the coating surface, the samples were treated as follows. Prepare 90% acetone solution, pour 45 mL into each test tube, add two drops of magnesium carbonate solution, place the rinsed and washed samples in, and extract chlorophyll for 24 h in a biochemical incubator in a dark environment at 8 °C. After completion, place 10 mL of the supernatant from each test tube into a centrifuge tube, with a centrifugation time of 15 min at a speed of 4000 rpm. Then, take 3 mL of the supernatant and place it into a cuvette, measure it using a UV-2000 UV-Vis spectrophotometer, and record the absorbance at the wavelengths of 750 nm, 663 nm, 645 nm, and 630 nm. According to Formula (5), calculate the chlorophyll a value. The diatom removal rate was similarly calculated based on the chlorophyll a concentration of the rinsed and washed samples with Formula (4).

$$\rho_a = 11.64 \times (OD_{663})\, 2.16 \times (OD_{645}) + 0.10 \times (OD_{630}) \tag{5}$$

3. Results

3.1. Molecular Structure

The infrared spectrum of the FPS and PDMS coating is shown in Figure 3. There was no peak during the range from 3650 to 3583 cm^{-1}. This indicates that the OH group in both hydroxyl-terminated fluoropolysiloxane and polydimethylsiloxane was completely exhausted by cross-linking reaction during the curing process. For the PDMS coating, the antisymmetric stretching vibration peak was located near 2962 cm^{-1}; the stretching vibration peak of the C-H bond in the CH$_3$ functional group could be seen near 2905 cm^{-1}; the symmetrical deformation vibration peak of the C-H bond appeared near 1257 cm^{-1}; the antisymmetric stretching vibration peak reflected by the Si-O-Si chemical bond of the siloxane main chain appeared near 1008 cm^{-1}; the characteristic peak of stretching vibration formed by the Si-O bond between the siloxane main chain and the branched-chain appeared near 785 cm^{-1}. The appearance of distinct peaks indicates that the silicone coating was prepared successfully. For the FPS coating, the above-mentioned peaks were observed, and they moved to left about 3 cm^{-1}. In addition, the absorption peak of CF$_3$ appeared near 1446 cm^{-1}. The characteristic peak of C-F deformation vibration appeared near 1205 cm^{-1} and 1060 cm^{-1}, and the Si-C stretching vibration peak appeared near 837 cm^{-1} and 764 cm^{-1}. The appearance of distinct peaks indicated that the fluorosilicone coating was successfully prepared.

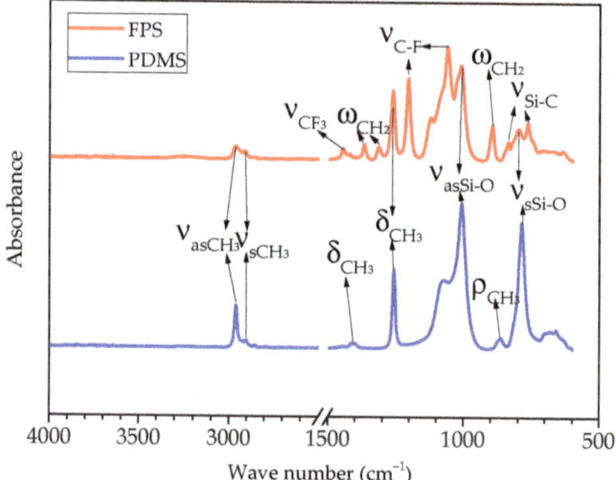

Figure 3. Infrared spectrum of studied coatings.

3.2. Surface Morphology and Roughness

The surface morphology and roughness of the studied coatings cured for 7 days in ambient are shown in Figure 4. For the FPS and PDMS coatings without silicone oil, their surfaces looked smooth and flat. However, the surface of the coatings containing silicone oil did not look smooth, and there were many circle ridges that leached silicone oil. The roughness of the coatings with leached silicone oil was obviously higher than those of the others. This phenomenon is similar to that reported in the research [28–30]. They have proven that this is because of the incompatibility between nonreactive silicone oil with low molecular weight and FPS or PDMS.

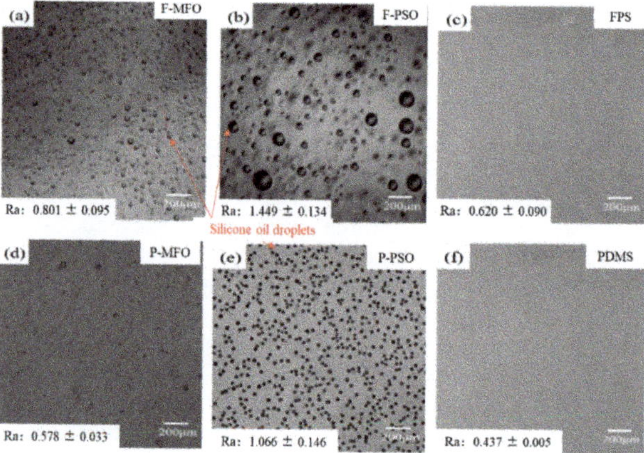

Figure 4. CLSM surface morphology and roughness of studied coatings: (**a**) F-MFO; (**b**) F-PSO; (**c**) FPS; (**d**) P-MFO; (**e**) P-PSO; (**f**) PDMS.

3.3. Interface Performance

The contact angle results of coatings are shown in Figure 5. The WCAs and DCAs of the FPS coatings are larger than those of the PDMS coatings. Especially, the DCAs of the FPS coatings are also larger than 90°, about 30 degrees higher than those of the PDMS coating. That indicated that the FPS coatings are not only hydrophobic but also oleophobic. The surface free energies of the FPS coatings, as shown in Figure 6, are 6.05 mJ/m^2, 9.23 mJ/m^2, and 10.19 mJ/m^2, and they are much lower than the surface energy of the PDMS coating (21.26 mJ/m^2, 22.53 mJ/m^2, and 24.95 mJ/m^2, respectively). This is because the outermost part of the fluorosilicone resin is densely filled with CF$_3$ groups, and the polarity of the F atom is significant, which combines with the flexibility of the main chain of the Si-O bond to improve the hydrophobicity and oil repellency of the coatings, and thus reduces the surface energy of the coating.

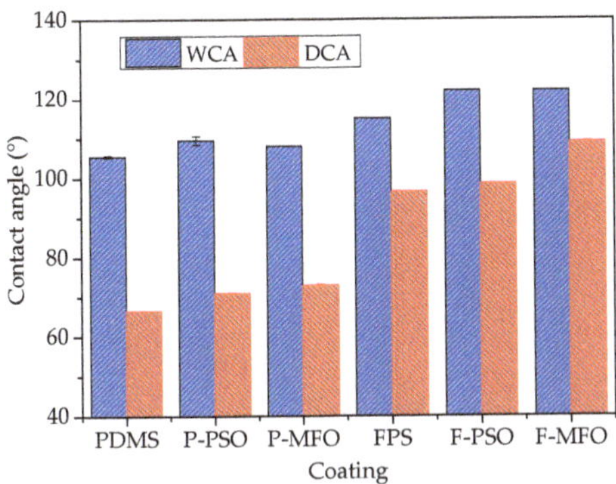

Figure 5. WCA and DCA of studied coatings.

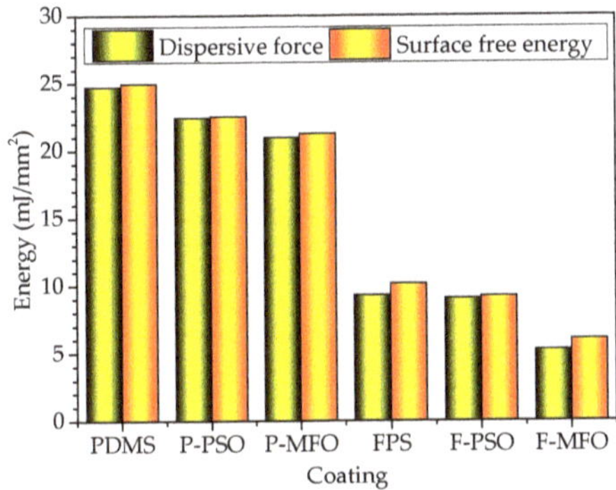

Figure 6. Surface energy and dispersive force of studied coatings.

3.4. Tensile Properties

The tensile curves and tensile properties of the studied coatings are shown in Figures 7 and 8. The fracture strength and the stress at 100% elongation of the three PDMS coatings are obviously higher than those of the three FPS coatings. This can be attributed to the strengthening effect of titanium dioxide being far inferior to that of fumed silica. The elastic modulus of the coatings without silicone oil is much higher than that of the ones containing silicone oil. Introducing silicone oil can result in the reduction in the elastic modulus. Among them, the elastic modulus of the F-PSO coating is least, 0.53 MPa. This is because during the curing process of the coating, the silicone oil does not participate in the reaction and is free between the long molecular chains, decreasing the elastic modulus and stretching strength.

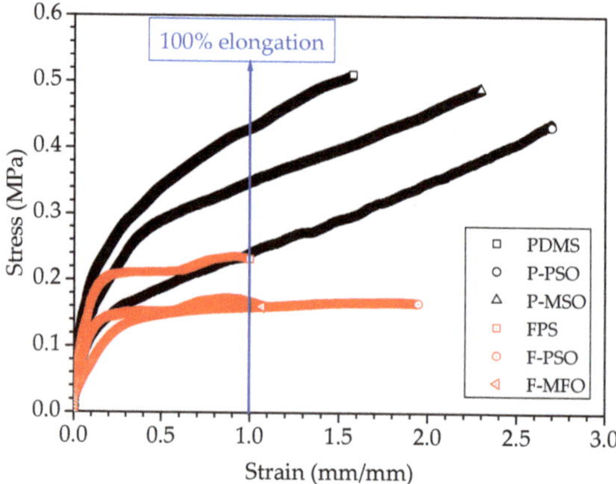

Figure 7. Tensile curves of studied coatings.

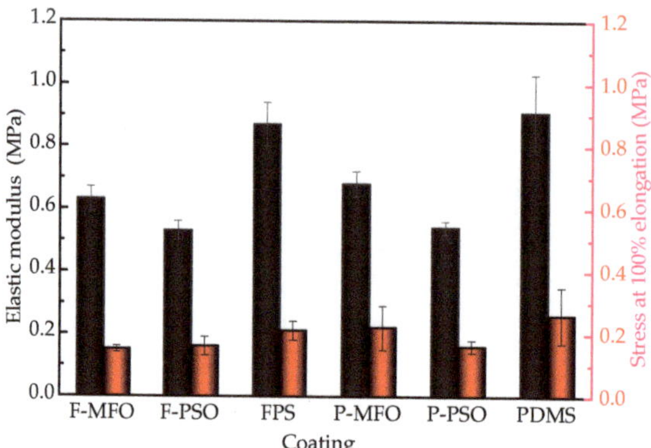

Figure 8. Elastic modulus and stress at 100% elongation of studied coatings.

3.5. Antifouling Performance

3.5.1. Anti-Bacterial Adhesion

Some images of marine bacterial colonies on the solid medium are shown in Figure 9. The colony concentration and removal rates of bacteria attached on the studied coatings are shown in Figures 10 and 11. Clearly, the colony concentration on washed samples is less than that on rinsed ones. In addition, that on the coatings containing silicone oil is obviously less than that on the coatings without silicone oil. The adhesive bacteria on the coatings containing silicone oil are more easily removed than that on the coatings without silicone oil. The removal rate of the coatings ranges from 50% to 80%. The removal rate of the FPS coatings is higher than that of the PDMS coatings, and that of the coatings containing silicone oil is higher than that of the ones without silicone oil. The highest is Coating F-PSO, which is 80.77%. The second is Coating F-MFO, which is 73.40%.

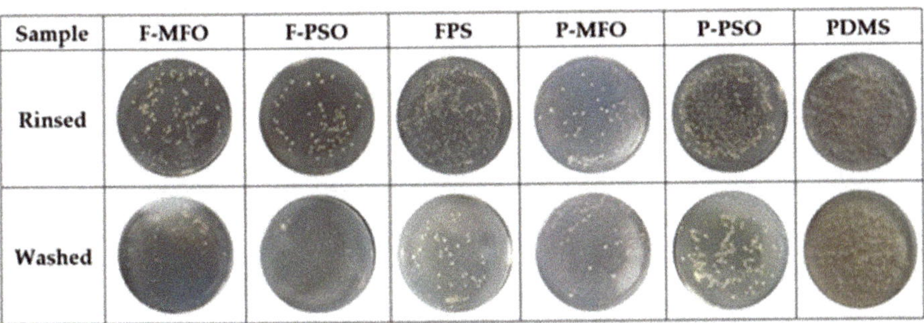

Figure 9. Original images of marine bacterial colonies.

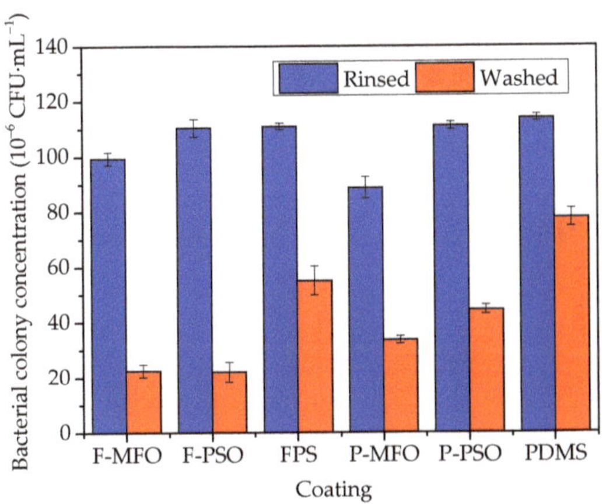

Figure 10. Colony concentration of bacteria attached on studied coatings.

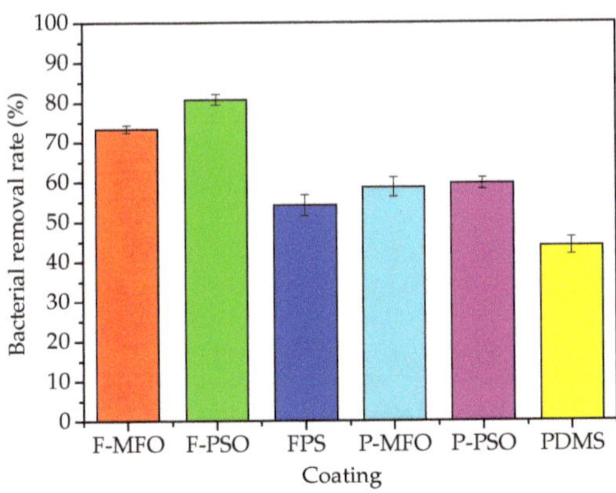

Figure 11. Removal rate of bacteria attached on studied coatings.

3.5.2. Anti-Diatom Adhesion

The Chlorophyll a concentration and removal rates of *Navicula Tenera* attached on the studied coatings are shown in Figures 12 and 13. Clearly, the concentration on the washed samples is less than that on the rinsed ones. In addition, that on the coatings containing silicone oil is obviously less than that on the coatings without silicone oil. The adhesive

Navicula Tenera on the coatings containing silicone oil is more easily removed than that on the coatings without silicone oil. The removal rate of the FPS coatings is higher than that of the PDMS coatings. Coating F-PSO has the highest diatom removal rate, which can reach 82.81%, followed by Coating F-MFO, which has a removal rate of 75.81%. Meanwhile, the FPS coating without silicone oil has a diatom removal rate of 52.46%, which is comparable to the diatom removal rate of the PDMS coating containing silicone oil.

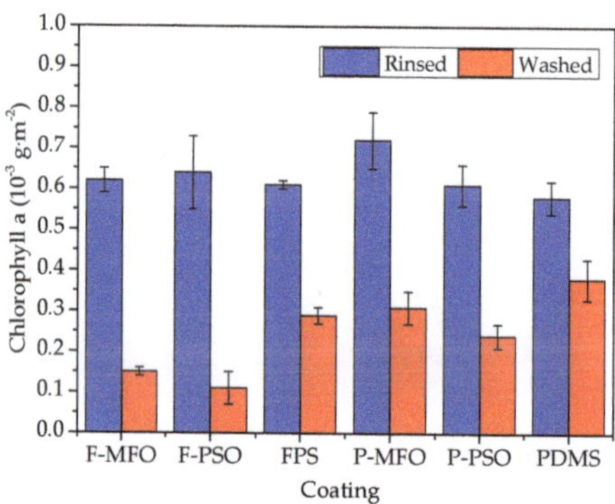

Figure 12. Chlorophyll a concentration of *Navicula Tenera* on studied coatings.

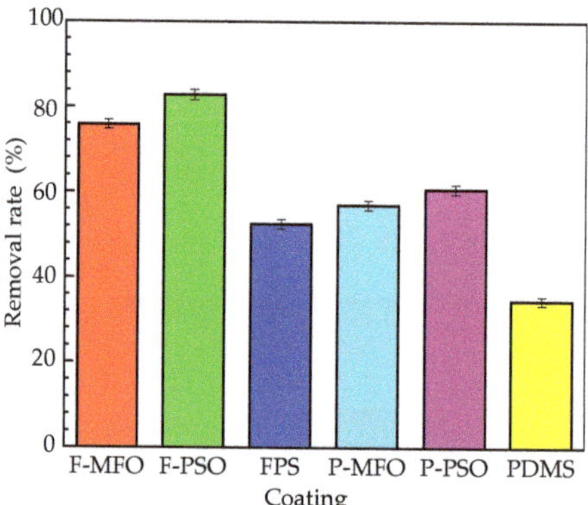

Figure 13. Removal rate of *Navicula Tenera* attached on studied coatings.

3.6. Discussion

Previous studies on fouling-release coatings have shown that the antifouling performance of coatings is related to various factors such as the surface energy, roughness, and elastic modulus of coatings [31–34]. Baier established the empirical relationship between biofouling adhesion and critical surface tension, the well-known Baier curve [3]. The surface morphology of the coating affects the attachment of marine organisms and the detachment of foulants, and the hydrophobic surface with a high roughness has a better interface performance [28]. The more the biofilms adhere to the washed samples, the worse the performance of the antifouling coating in the actual use of the marine environment [30]. Researchers have found that the wettability [31], surface chemistry [32], and surface lubricity [33] of the coating surface have a significant effect on diatom adhesion [34]. Brady [35] proposed that the relative adhesion of silicone fouling-release coatings is to the power of 1/2 of the product of elastic modulus E and surface free energy γ, i.e., Brady's equation $Ar = \sqrt{E\gamma}$. It is considered as an important indicator for selecting and evaluating fouling-release coatings.

In this study, whether for bacteria or *Navicula Tenera*, its removal rate does not simply decrease with the increase in the surface energy, as shown in Figure 14. However, the removal rate of the FPS coatings is much larger than that of the PDMS coatings. Whether for FPS or PDMS coatings, the removal rate of the coatings containing silicone oil is much larger than that of the coatings without silicone oil. This is because the nonreactive low-molecular-weight silicone oil is incompatible with the matrix resin and migrates from the inside to the surface, forming a lubricating surface that increases the smoothness and hydrophobicity of the coating and decreases the surface energy [30], thereby improving the coating antifouling performance. Fluorosilicone polymer has a particular polarity and low surface energy, which can remove foulants by shearing between interfaces, and the CF_3 group in the FPS coatings promotes the antifouling performance of the coatings. In addition, whether it is an FPS or PDMS coating, the coatings containing PSO have a higher removal rate of adhesion biofouling than that containing MFO. This is because the compatibility of PSO and the matrix resin FPS or PDMS is lower than that of MFO and the matrix resin FPS or PDMS, and phase separation is more likely to occur; PSO is also easier to migrate to the coating surface than MFO is. As shown in Figure 4, more and larger PSO droplets are observed than MFO droplets, which weakens the binding force between the attached organisms and the coating. Therefore, the attached organisms are easier to remove by the seawater washing, and the removal rate of the coating F-PSO is higher than that of the coating F-MFO.

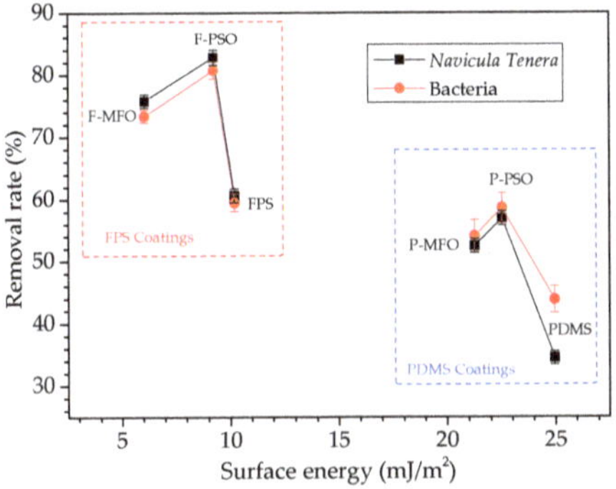

Figure 14. Surface energy vs. biofouling removal rate on studied coatings.

According to Brady's equation, the removal rates of biofouling and the relative adhesion of the studied coatings are reported in Figure 15. Obviously, whether for FPS or PDMS coatings, both the marine bacteria and *Navicula Tenera*, its removal rate linearly increases with the decrease in Brady's relative adhesion.

In order to better compare the antifouling properties of coatings, the relative ratio of the colony concentration of bacteria attached and chlorophyll a concentration of *Navicula Tenera* on each washed coating to the washed PDMS coating was calculated and is reported in Figure 16. Clearly, whether for bacteria or *Navicula Tenera*, the relative ratio of the FPS coatings is much less than that of the PDMS coatings. Whether for FPS or PDMS coatings, the relative ratio of the coatings containing silicone oil is much less than that of the coatings without silicone oil. Consequently, the prepared fluorosilicone fouling-release coatings showed much better antifouling performance than PDMS coatings in this study and good application prospects. Furthermore, introducing nonreactive silicone oil into PDMS or FPS coatings can further improve the antifouling performance of the coating to a certain extent.

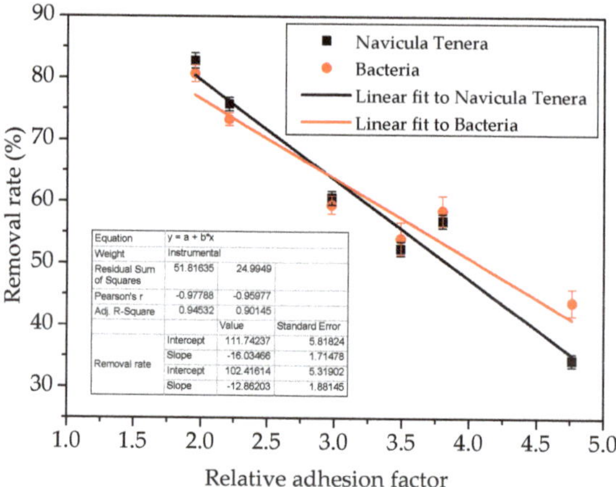

Figure 15. Removal rate of biofouling and relative adhesion of studied coatings.

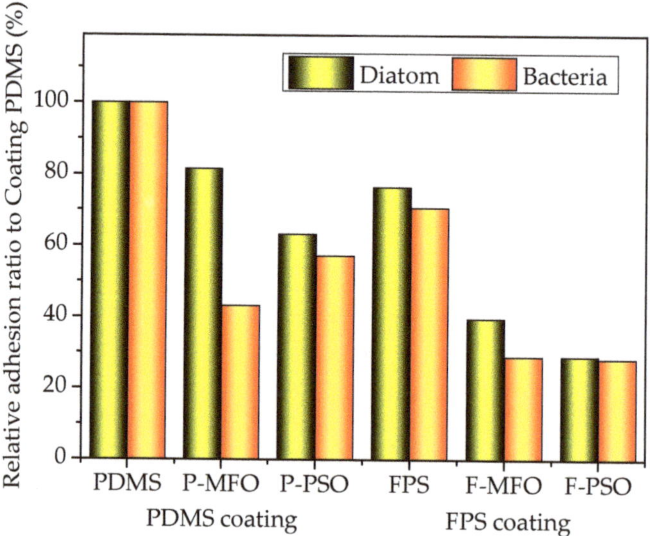

Figure 16. Relative ratio of attached organism on studied coating to PDMS coating.

4. Conclusions

In this paper, fluorosilicone coatings (FPS) were prepared using hydroxyl-terminated fluoropolysiloxane as the primary film-forming material, TEOS as the curing agent, and DBTDL as the catalyst. The FPS coatings have a larger WCA and smaller surface free energy than that of the PDMS coatings. The FPS coatings exhibited better hydrophobic and oleophobic properties. The elastic modulus and tensile strength of the FPS coatings are lower than those of the PDMS coatings. The FPS coatings have better antifouling performance than the PDMS coatings. The prepared fluorosilicone fouling-release coatings show good application prospects. Introducing nonreactive silicone oil into PDMS or FPS coatings can improve the antifouling performance of the coating to a certain extent.

Author Contributions: T.W., Z.Z. and Y.Q., conceptualization; T.W., Z.Z. and Y.Q., methodology; T.W., software; T.W., validation; T.W. and Z.Z., formal analysis; T.W., Q.C. and C.G., investigation; Y.Q. and Z.Z., resources; T.W., data curation; T.W., writing—original draft preparation; T.W. and Z.Z., writing—review and editing; T.W. and C.G., visualization; Z.Z. and Y.Q., supervision; Z.Z. and Y.Q., project administration; Z.Z. and Y.Q., funding acquisition. All authors have read and agreed to the published version of the manuscript.

Funding: This research was funded by the National Natural Science Foundation of China, grant number 51879021.

Institutional Review Board Statement: Not applicable.

Informed Consent Statement: Not applicable.

Data Availability Statement: The data in this study are contained within the article.

Conflicts of Interest: The authors declare no conflict of interest.

References

1. Schultz, M.P. Effects of coating roughness and biofouling on ship resistance and powering. *Biofouling* **2007**, *23*, 331–341. [CrossRef] [PubMed]
2. Lejars, M.; Margaillan, A.; Bressy, C. Fouling release coatings: A nontoxic alternative to biocidal antifouling coatings. *Chem. Rev.* **2012**, *112*, 4347–4390. [CrossRef] [PubMed]
3. Jin, H.; Tian, L.; Bing, W.; Zhao, J.; Ren, L. Bioinspired marine antifouling coatings: Status, prospects, and future. *Prog. Mater. Sci.* **2021**, *124*, 100889. [CrossRef]
4. Su, X.; Yang, M.; Hao, D.; Guo, X.; Jiang, L. Marine antifouling coatings with surface topographies triggered by phase segregation. *J. Colloid Interface Sci.* **2021**, *598*, 104–112. [CrossRef]
5. Rilov, G.; Crooks, J.A. *Biological Invasions in Marine Ecosystems*; Springer: Berlin/Heidelberg, Germany, 2009. [CrossRef]
6. Fusetani, N. Antifouling marine natural products. *Nat. Prod. Rep.* **2011**, *28*, 400–410. [CrossRef]
7. Thomas, K.V.; Brooks, S. The environmental fate and effects of antifouling paint biocides. *Biofouling* **2010**, *26*, 73–88. [CrossRef]
8. Long, L.; Wang, R.; Chiang, H.Y.; Ding, W.; Li, Y.-X.; Chen, F.; Qian, P.-Y. Discovery of Antibiofilm Activity of Elasnin against Marine Biofilms and Its Application in the Marine Antifouling Coatings. *Mar. Drugs* **2021**, *19*, 19. [CrossRef]
9. Kristensen, J.B.; Meyer, R.L.; Laursen, B.S.; Shipovskov, S.; Besenbacher, F.; Horsmans Poulsen, C. Antifouling enzymes and the biochemistry of marine settlement. *Biotechnol. Adv.* **2008**, *26*, 471–481. [CrossRef]
10. Genzer, J.; Efimenko, K. Recent developments in superhydrophobic surfaces and their relevance to marine fouling: A review. *Biofouling* **2006**, *22*, 339–360. [CrossRef]
11. Hearin, J.; Hunsucker, K.Z.; Swain, G.; Stephens, A.; Gardner, H.; Lieberman, K.; Harper, M. Analysis of long-term mechanical grooming on large-scale test panels coated with an antifouling and a fouling-release coating. *Biofouling* **2015**, *31*, 625–638. [CrossRef]
12. Xie, Q.; Pan, J.; Ma, C.; Zhang, G. Dynamic surface antifouling: Mechanism and systems. *Soft Matter* **2019**, *15*, 1087–1107. [CrossRef] [PubMed]
13. Wu, W.; Zhao, W.; Wu, Y.; Zhou, C.; Li, L.; Liu, Z.; Dong, J.; Zhou, K. Antibacterial behaviors of Cu_2O particles with controllable morphologies in acrylic coatings. *Appl. Surf. Sci.* **2019**, *465*, 279–287. [CrossRef]
14. Gudipati, C.S.; Greenlief, C.M.; Johnson, J.A.; Prayongpan, P.; Wooley, K.L. Hyperbranched fluoropolymer and linear poly (ethylene glycol) based am-phiphilic crosslinked networks as efficient antifouling coatings: An insight into the surface compositions, topographies, and morphologies. *J. Polym. Sci. Part A Polym. Chem.* **2004**, *42*, 6193–6208. [CrossRef]
15. Xie, S.; Wang, J.; Wang, L.; Sun, W.; Lu, Z.; Liu, G.; Hou, B. Fluorinated diols modified polythiourethane copolymer for marine antifouling coatings. *Prog. Organ. Coat.* **2020**, *146*, 105733. [CrossRef]
16. Wang, X.; Yu, L.; Liu, Y.; Jiang, X. Synthesis and fouling resistance of capsaicin derivatives containing amide groups. *Sci. Total Environ.* **2020**, *710*, 136361. [CrossRef]
17. Zhou, W.M. *Research on Marine Antifouling Performances of Biologic Surface*; National University of Defense Technology: Changsha, China, 2010.
18. Brady, R.F. Properties which influence marine fouling resistance in polymers containing silicon and fluorine. *Prog. Organ. Coat.* **1999**, *35*, 31–35. [CrossRef]
19. Wu, J.Y. *The synthesis, Vulcanizaayion and Modification of Fluorinated Polysiloxanes*; Beijing University of Chemical Technology: Beijing, China, 2017.
20. Martinelli, E.; Hill, S.D.; Finlay, J.A.; Callow, M.E.; Callow, J.A.; Glisenti, A.; Galli, G. Amphiphilic modified-styrene copolymer films: Antifouling/fouling release proper-ties against the green alga *Ulva linza*. *Prog. Organ. Coat.* **2016**, *90*, 235–242. [CrossRef]
21. Sun, Q.; Li, H.; Xian, C.; Yang, Y.; Song, Y.; Cong, P. Mimetic marine antifouling films based on fluorine-containing polymethacrylates. *Appl. Surf. Sci.* **2015**, *344*, 17–26. [CrossRef]
22. Mera, A.E.; Wynne, K.J. Fluorinated Silicone Resin Fouling Release Composition. Google Patents US6265515B1, 24 July 2001.

23. Stafslien, S.; Daniels, J.; Mayo, B.; Christianson, D.; Chisholm, B.; Ekin, A.; Webster, D.; Swain, G. Combinatorial materials research applied to the development of new surface coatings IV. A high-throughput bacterial biofilm retention and retraction assay for screening fouling-release performance of coatings. *Biofouling* **2007**, *23*, 45–54. [CrossRef]
24. Owens, D.K.; Wendt, R.C. Estimation of the surface free energy of polymers. *J. Appl. Polym. Sci.* **1969**, *13*, 1741–1747. [CrossRef]
25. *GB/T 528-2009/ISO 37:2005*; Rubber, Vulcanized or Thermoplastic-Determination of tensile Stress-Strain Properties. Standards Press of China: Beijing, China, 2009. (In Chinese)
26. Zhang, Y.; Qi, Y.; Zhang, Z. Synthesis of PPG-TDI-BDO polyurethane and the influence of hard segment content on its structure and antifouling properties. *Prog. Organ. Coat.* **2016**, *97*, 115–121. [CrossRef]
27. Liu, H.; Zhang, Z.P.; Qi, Y.H.; Liu, S.; Lin, J.; Fang, K.; Zhang, Z. Experimental methods of evaluating biofouling of marine benthic diatoms on non-toxic antifouling coating. *Mar. Environ. Sci.* **2006**, *25*, 89–92.
28. Ba, M.; Zhang, Z.; Qi, Y.-h. The leaching behavior of phenylmethylsilicone oil and antifouling performance in nano-zinc oxide reinforced phenylmethylsilicone oil–Polydimethylsiloxane blend coating. *Prog. Organ. Coat.* **2018**, *125*, 167–176. [CrossRef]
29. Ba, M.; Zhang, Z.-P.; Qi, Y.-H. The influence of MWCNTs-OH on the properties of the fouling release coatings based on polydimethylsiloxane with the incorporation of phenylmethylsilicone oil. *Prog. Org. Coat.* **2019**, *130*, 132–143. [CrossRef]
30. Jiang, Y.; Zhang, Z.; Qi, Y. The Compatibility of Three Silicone Oils with Polydimethylsiloxane and the Microstructure and Properties of Their Composite Coatings. *Polymers* **2021**, *13*, 2355. [CrossRef]
31. Compere, C.; Bellon-Fontaine, M.N.; Bertrand, P.; Costa, D.; Marcus, P.; Poleunis, C.; Pradier, C.-M.; Rondot, B.; Walls, M.G. Kinetics of conditioning layer formation on stainless steel immersed in seawater. *Biofouling* **2001**, *17*, 129–145. [CrossRef]
32. Finlay, J.A.; Callow, M.E.; Ista, L.K.; Lopez, G.L.; Callow, J.A. The influence of surface wettability on the adhesion strength of settled spores of the green alga Enteromorpha and the diatom Amphora. *Integr. Comp. Biol.* **2002**, *42*, 1116–1122. [CrossRef]
33. Bowen, J.; Pettitt, M.E.; Kendall, K.; Leggett, G.J.; Preece, J.A.; Callow, M.E.; Callow, J.A. The influence of surface lubricity on the adhesion of Navicula perminuta and Ulva linza to alkanethiol self-assembled monolayers. *J. R. Soc. Interface* **2007**, *4*, 473–477. [CrossRef]
34. Li, Y.; Gao, Y.H.; Li, X.S.; Yang, J.Y.; Que, G.H. Influence of surface free energy on the adhesion of marine benthic diatom Nitzschia closterium MMDL533. *Colloids Surf. B Biointerfaces* **2010**, *75*, 550–556. [CrossRef]
35. Brady, R.F., Jr.; Singer, I.L. Mechanical factors favoring release from fouling release coatings. *Biofouling* **2000**, *15*, 73–81. [CrossRef]

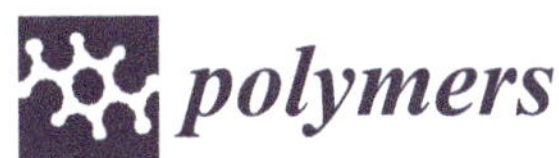 

Article

New Insight on Photoisomerization Kinetics of Photo-Switchable Thin Films Based on Azobenzene/Graphene Hybrid Additives in Polyethylene Oxide

Qais M. Al-Bataineh [1], Ahmad A. Ahmad [1,*], Ahmad M. Alsaad [1] and Ahmad Telfah [2,3]

[1] Department of Physics, Jordan University of Science & Technology, P.O. Box 3030, Irbid 22110, Jordan; qalbataineh@ymail.com (Q.M.A.-B.); alsaad11@just.edu.jo (A.M.A.)

[2] Leibniz Institut für Analytische Wissenschaften-ISAS-e.V., Bunsen-Kirchhoff-Straße 11, 44139 Dortmund, Germany; telfah.ahmad@isas.de

[3] Hamdi Mango Center for Scientific Research (HMCSR), The Jordan University, Amman 11942, Jordan

* Correspondence: sema@just.edu.jo or sema_just@yahoo.com

Received: 14 November 2020; Accepted: 7 December 2020; Published: 10 December 2020

Abstract: In this work, we reported a new insight on the kinetics of photoisomerization and time evolution of hybrid thin films considering the azo-dye methyl red (MR) incorporated with graphene accommodated in polyethylene oxide (PEO). The kinetics of photoisomerization and time-evolution of hybrid thin films were investigated using UV-Vis s and FTIR spectroscopies, as well as appropriate models developed with new analytical methods. The existence of azo-dye MR in the complex is crucial for the resource action of the *trans* ↔ *cis* cycles through UV-illumination ↔ Visible-illumination relaxations. The results of the UV–Vis and the FTIR investigations prove the cyclical *trans* ↔ *cis*-states. Consequently, PEO-(MR-Graphene) hybrid composite thin films can be introduced as possible applicants for photochromic molecular switches, light-gated transistors, and molecular solar thermal energy storage media.

Keywords: polymerized nanocomposite thin films; photoisomerization processes; trans-cis-isomers; molecular solar thermal energy storage media; photo-switchable thin films

1. Introduction

Photo-switchable thin films can be isomerized between two metastable states through light-illumination. This type of thin-film has gained noticeable attention in many applications in physics, chemistry, and biology. The striking feature of photo-switchable thin films is that the two isomers have different physical and chemical properties. Recently, the azobenzene (AZO) has gained great momentum owing to its potential applications in photochromic molecular switches [1], light-gated transistors [2], and molecular solar thermal energy storage media [3–6]. AZO has two geometric forms, *trans*-state and *cis*-state. All AZO thin films are initially in the *trans*- isomerization state since it is thermally stable at room temperature [7,8]. The *trans* → *cis* isomerization may occur by exposing thin films to the UV-light illumination, while the reverse *cis* → *trans* isomerization can occur by either illuminating thin films with visible light (400–450 nm) or by thermal excitation in a dark environment [9,10]. The photoisomerization mechanisms of the *trans*↔*cis* isomers in the blended polymers play a key role in extracting the proper chemo physical properties of additive azo benzene [7,10–13].

Polymers go under a photodegradation process via UV light-absorption (in particular) by backbone-carbonyl groups induced by photochemical reactions. Photodegradation of polymers usually occurs through either chain scission (reduction in molecular weight) formatting C=C double bonds, cross-linking, and hydroperoxide O–H reactions that activate the polymer-molecule while absorbing light. The degradation process begins with light-absorption by a photo-initiator, then photo-cleavage of polymeric-molecules, transferring them into free-radicals that further enhances the degradation process. The level of degradation (breakage) is relevant to the light-energy used. Shorter light-wavelengths do not need oxygen for the hydroperoxide process, while low-wavelength sources need an oxygen environment [14–17]. Azo dyes usually degrade under the influence of UV-irradiation, especially when a photocatalytic source of metal oxide is presented. ZnO, TiO2, SnO2, and CuO are mostly used as photocatalytic oxidation sources for the degradation process [18,19]. Methyl red is of those azo dyes which contain one or more azo groups (–N=N–) as a chromophore group that is influenced by UV absorption. Recently, 2D carbon as graphene and graphene oxides were used as support materials for photocatalytic processes [20,21].

The mutuality of the exceptional transformation of the photoisomerization processes via thermally relaxed stable *trans*-isomerization and nonstable *cis*-isomerization via illumination has become of the main theoretical and practical perspectives [22–24]. Generally, the photoisomerization process exhibits several mutual operations involving bending or non-bending aromatic rings in micromolecular composite chains [25,26].

We selected polyethylene oxide (PEO) as a host polymer owing to its semi-crystallinity nature with two phases of crystallite and amorphous forms coexisting at normal conditions [27] and its low absorbance values compared to the other host polymers [28–30]. Graphene has flattened sp^2-crossed networks with π-electron restrained over the rings [31]. Graphene integrates groups of carboxylic, carbonyl, and hydroxyl assemblies at the boundaries, including hydroxyl and epoxy assemblies at the basal planes. In addition, the aromatic AZO acts as a frail electrophile and outbreaks carbon atoms in a dense electron fog form in a hydroxybenzene ring [32]. One of the important benefits of selecting graphene in the composite is its behavior as a photo-initiator [33].

This work's main objective is to investigate the kinetics of photoisomerization processes of polymeric thin films performed with azo-dye (methyl red) incorporated with graphene accommodated in polyethylene oxide (PEO). Understanding the mechanism of photoisomerization is crucial for the practical implementation of this nanocomposite in device fabrication. The optical and chemical characteristics of the *trans-* and *cis*-isomers of the PEO-MR-Graphene hybrid complexes are usually investigated via the UV–Vis and the FTIR techniques. To the best of our knowledge, we are not aware of any practical investigation of the kinetics of PEO-MR-Graphene hybrid composites' methodological isomerization via optical studies.

2. Experimental Details

2.1. Materials

Polyethylene Oxide (PEO) (–CH_2CH_2O–)$_n$ with a molecular weight of 300,000 g/mol and graphene powder (<20 μm) with a molecular weight of 12.01 g/mol were obtained from Sigma-Aldrich Co. Inc: Munich, Germany Methyl-Red (MR) ($C_{15}H_{15}N_3O_3$) of pH level between 4.2 and 4.6 powder supplied by SCP SCIENCE: Montreal, QC, Canada.

2.2. Synthesis of PEO-MR-Graphene Hybrid Composite Thin Films

All solutions were prepared using absolute methanol (CH_3OH, with a purity of 99.8%). One gram of PEO was successively dissolved in 100 mL methanol at 45 °C by using continuous magnetic stirring for 5 h. MR-Graphene solution (1 mole: 1 mole) was mixed via solid-state blending using an agate mortar (BUCHI™ Achat Mörser mit Pistill: Fisher scientific, Hampton, NH, United States). After that, one gram of MR-Graphene mixture was dissolved in 100 mL methanol at room temperature. The hybrid solution of PEO-(MR-Graphene)/methanol was obtained by mixing PEO/methanol and MR-Graphene/methanol in a (3 to 1) ratio via stirring them magnetically for a duration of 6 h.

The composites were homogenized via sonication for 6 h. The final solution was treated with a 0.45 µm Millipore filter. The hybrid thin films with 300 nm thickness (measured in cross-sectional view using a SEM micrograph: Fisher scientific, Hampton, NH, United States) were dip-coated for one hour on glass substrates. The solvents were evaporated, and the organic residues were removed by air drying the films at 70 °C for half an hour.

2.3. Characterization

The absorbance measurements were performed by a Double-Beam UV–vis Spectrophotometer (U-3900H:Hitachi, Tokyo, Japan) at room temperature. The hybrid nanocomposite vibrational bands were obtained using the Fourier transform infrared spectroscopy (FTIR) (Bruker Tensor 27 spectrometer with a disc of KBr: Billerica, MA, United States spectrometer with a disc of KBr). FTIR measurements were performed by pealing-off the films out of the glass substrates and used as a solid form in the system. The thermal stability was investigated using the Thermogravimetric Analysis (TGA) technique.

The PEO-(MR-Graphene) thin-film were illuminated by a UV-light source with 366 nm wavelength and 6 Watts of power (32 W/cm^2 of intensity) for 0, 30, 60, 120, 240, and 480 s in order to investigate the impact of the time-evolved in transforming the *trans*-isomerization phase to *cis*-phase. The absorbance spectra were measured at every UV-illumination time-exposure. The films were exposed to blue light-illumination by a visible source with a wavelength of 467 nm and power of 6 Watts (32 W/cm^2 of intensity) for 0, 90, 180, 360, 720, 1440 s. The process was done to inspect the influence of the exposure-time on the reversal of the monomer *cis*-isomerization state back to the original *trans*-state. The absorbance spectra were measured at every exposure-time evolved by the visible-illumination. Moreover, the FTIR spectroscopy measurements were employed as additional supportive evidence of the influence of the UV-Vis illuminations via exploring the vibrational changes that occurred in the bonding modes of the *trans*- and the *cis*-cases, accordingly.

3. Results and Discussion

3.1. Characterization of PEO-(MR-Graphene) Hybrid Nanocomposite Thin Film

As previously reported [3,33], the chemical structure of the AZO-Graphene used in this work can be visualized, as shown in Figure 1a. The coupling agent forming the AZO-Graphene hybrid contains the AZO-nitrogen atom bonded covalently with the carbon atom that is also bonded with the graphene aromatic ring. The chemical formula of PEO is $C_{2n}H_{4n+2}O_{n+1}$, and that of MR is given by $C_{15}H_{15}N_3O_3$. The covalent linkage between methyl red and Graphene exterior superficial shells can be determined by FTIR patterns shown in Figure 1b. The vibrational bands for graphene located at 2015 cm^{-1} and 2161 cm^{-1} represent C=C and C=O. FTIR spectrum for PEO exhibited vibrational bands between 700–1000 cm^{-1} representing bending C–H. Vibrational bands between 1000–1400 cm^{-1} are assigned to C–O stretching vibration, a band at 1470 cm^{-1} that could be ascribed to the –CH$_2$ bending vibration, while a band appearing at the 2886 cm^{-1} could be ascribed to the symmetric and asymmetric C–H stretching modes of the CH$_2$ group. Moreover, the significant pronounced vibrational bonds that appeared in PEO-(MR–Graphene) hybrid nanocomposite thin films were found to be related to the aromatic rings, the azo- chromophores (–N=N–), the stretching bonds of C–N, and others such as C–H. The in-plane C–H vibrational peak was found at 1122 cm^{-1}, and the out-of-plane vibrational peaks were found between 1000–700 cm^{-1}. The aromatic bands were observed between 1600 and 1400 cm^{-1}. Moreover, the spectral peak observed at 1342 cm^{-1} was apportioned to the C–O stretching bond in PEO. According to W. Pang et al. study [3], the absorption due to the C–N bonding is unlike the C=C bonding, and its frequency-range is centered at 1342 cm^{-1} caused by the influence of resonance which upsurges the bond-order assigned to that particular ring in the chain and the dangling N-atom. The presence of C–N bonds designates the covalent bonding between MR to the graphene-outer surface. The peak observed at 3732 cm^{-1} was assigned to the O–H bond in PEO. However, the peak at 2861 cm^{-1} was dispensed to C–H stretching. Moreover, the peak found at 2114 cm^{-1} was assigned to double bonding C=C. The peaks observed at 1692 cm^{-1} and 1583 cm^{-1} were related to the double bonding of C=O. Finally, the peak observed at 3732 cm^{-1} was consigned to the single bonding of O–H.

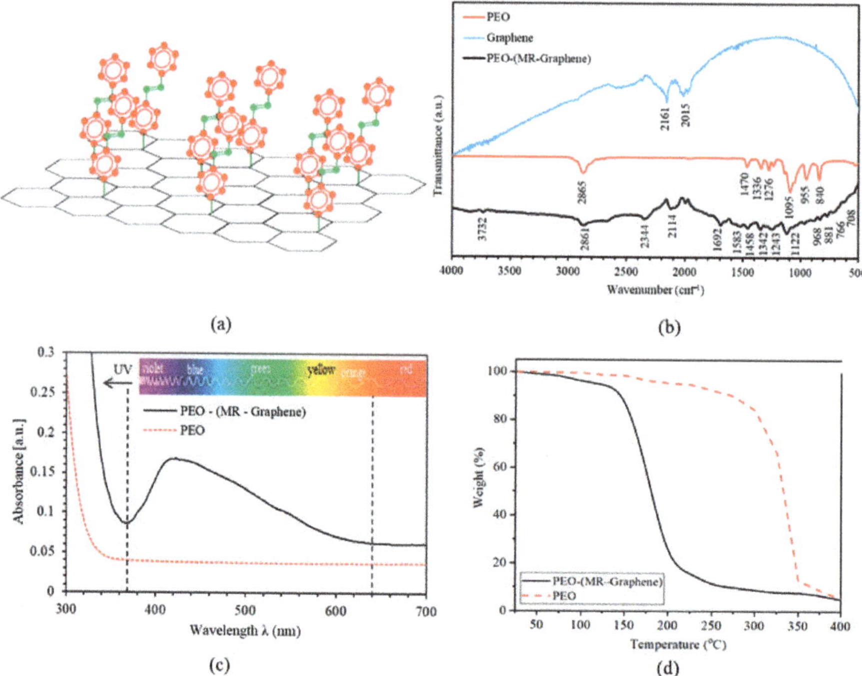

Figure 1. (**a**) The molecular structures of *trans*-AZO-Graphene hybrid, (**b**) FTIR spectra of PEO, Graphene and PEO-(MR–Graphene) hybrid composite thin film, (**c**) Absorbance spectra of PEO and PEO-(MR–Graphene) hybrid composite thin films versus the wavelength, (**d**) TGA curve for PEO-(MR–Graphene) hybrid composite thin film.

Figure 1c shows PEO and PEO's absorbance spectra-(MR–Graphene) hybrid composite thin films as a function of wavelength. The PEO absorption spectra-(MR–Graphene) hybrid composite is characterized by a band transition. Namely, the $\pi - \pi^*$ in the range of 350–650 nm. However, PEO thin film does not contain any absorption band identified in this region [9]. Clearly, the absorbance of the PEO-(MR–Graphene) hybrid composite thin film exhibits $\pi - \pi^*$ transition in the (350–650) nm range with a maximum at 422.5 nm. This spectral behavior is believed to occur due to the presence of MR in the film nanocomposite giving the films the red-appearance color. Hence, the films show high absorbance for green, blue, and violet-colored light [7,10–13].

Furthermore, PEO's thermal stability-(MR–Graphene) hybrid nanocomposites were investigated using the thermogravimetric analysis (TGA) technique at temperatures up to 400 °C, as shown in Figure 1d. The PEO-(MR–Graphene) hybrid nanocomposite shows a stable weight-loss starting at the ambient temperature T and extends to 150 °C, at which the mass-loss becomes 8% approximately. The mass-loss was considered as the loss of water/solvents adsorbed in the sample. The mass-loss curve versus temperature then sharply declined in the temperature interval ranging between 150 to 200 °C. Within this interval, the mass-loss is significant (80%), indicating the influence of the intermolecular/intramolecular bonding change when the nanocomposite is exposed to high temperatures. PEO's thermal stability-(MR-Graphene) hybrid nanocomposite is lower than that of PEO, as shown in Figure 1d. For PEO, the mass-loss versus temperature drops sharply at around 300 °C [34,35]. Interestingly, despite the slight negligible slope in the TGA relation below 150 °C,

TGA thermograms confirm that the PEO-(MR–Graphene) hybrid composite was stable below 150 °C. Most of the practical optical applications are feasible below 150 °C.

3.2. Kinetics of Photoisomerization Processes

In this section, we investigated and reported the kinetics of PEO transformation-(MR–Graphene) hybrid nanocomposite thin films from the original *trans*-state to *cis*-state using UV-exposure. The reversed transformation to *trans*-state via visible-illumination was also discussed and interpreted. UV–light primarily illuminates the PEO-(MR–Graphene) hybrid nanocomposite thin films for a certain period of time. The films are then exposed to Blue-light illumination for another period. Figure 2a illustrates the absorbance spectra of the PEO-(MR–Graphene) hybrid nanocomposite thin films illuminated by UV-light for various periods. The major absorption peak of PEO-(MR–Graphene) at the initial *trans*-state in the visible range was found to be at 422.5 nm with an absorbance amplitude of 0.169%. The film was then exposed to UV light of 366 nm for 30, 60, 120, 240, and 480 s, respectively. Moreover, the films show a variant absorbance band in the middle of the visible region (380–625 nm) with a peak-blue-shifted and amplitude-decrease transferring the *trans*-state to a *cis*-state, as expected. The photoisomerization process is not a spontaneous instantaneous process. Instead, it occurs through steadily building up sub-stages. The film exposure to the UV-illumination reveals a noticeable reduction in the absorbance at ~422.5 nm in the *trans*-isomerization stage and the presence of a dual peak at ~470 nm created from the *cis*-isomerization state. A steady reduction in the π–π^* electron transition band of the *trans*-state upon the increasing of UV-illumination time duration until a photostationary equilibrium state between the *trans*- and the *cis*-isomers was detected. The absorbance spectra of PEO-(MR–Graphene) thin films prior-to and in post-UV illumination revealed the existence of dual isosbestic figure-tips that appeared close to 380 and 625 nm, respectively. Moreover, the UV-illumination of PEO-(MR–Graphene) hybrid nanocomposite thin film leads to the transformation from *trans*-state to a *cis*-state. In other words, the absorbance peak at wavelengths ranging from 380 nm to 625 nm decreased as the period of illumination increased. The prolonged illumination does not lead to any changes in the absorption spectra, confirming that a quasi-state of photo-stationary phase occurs between the *trans*- and the *cis*-stats. The long-time isomerization (480 s of illumination) is needed until a photo-stationary equilibrium is achieved, compared with previous results reported [7,9,10]. The observed electronic transitions in the visible range indicated a suitably effective absorption of solar power with high-energy. The solar energy absorbed in the material is of prime necessity to achieve two- or three-multiples of energy more than the activation barrier-energy ΔE_i (needed for complete isomerization of AZO-molecules) [36]. Based on the findings in the literature, the energy reserved/AZO-molecules (ΔH) for *cis*-isomerization has to be less than ΔE_i. In other words, $\Delta H = \Delta E_i - \Delta E_a$, where, ΔE_i is the activation barrier (energy needed to isomerize one AZO molecule), ΔE_a is the thermal/optical activation barrier and efficiency of storing energy (the energy stored per the solar energy absorbed) is normally < 30% [36,37].

The kinetics of the *trans-cis* photoisomerization of PEO-(MR–Graphene) hybrid nanocomposite thin films were investigated using the first-order kinetics via calculating the rate of photoisomerization, as well as the thermal/optical activation barrier ΔE_a for *trans-cis* photoisomerization [38,39]. The photoionization rate as a function of time (p) could be written as,

$$\ln \frac{A_t - A_\infty}{A_0 - A_\infty} = -pt \tag{1}$$

where A_0, A_t and A_∞ are the absorbance at various time conducts, namely prior to the light exposure (initial *trans*-state), during radiation in time t, and post to the light exposure for a protracted time. The average value of $\ln(A_t - A_\infty / A_0 - A_\infty)$ of PEO-(MR–Graphene) hybrid nanocomposite thin film is plotted against the period (t), as shown in Figure 2b. Obviously, $\ln(A_t - A_\infty / A_0 - A_\infty) - t$ relation has two distinct linarites. The discontinuity occurs at $t_c \sim 240$ min, then the curve falls off with less steepness as a function of illumination time. The area's ratio under the absorption curve in the visible range for all UV-illumination periods for the *cis*-hybrid isomerization was calculated with respect to the trans-hybrid isomerization. It was found to be around 95.5% at t_c for the PEO-(MR–Graphene) hybrid nanocomposite thin films. The rate constant (p) was found from the slope to be $p_1 = 3.822 \times 10^{-3}$ s^{-1} for $0 < t < t_c$, $p_2 = 1.634 \times 10^{-4}$ s^{-1} for $t_\infty > t > t_c$. Moreover, the variations in p could be reasonably attributed to the disparity of ΔE_a [37] and calculated as,

$$\Delta E_a = -kT \ln \frac{h \ln 2}{\tau_{1/2} kT} \tag{2}$$

where: $\tau_{1/2}$ is the time needed to transfer half the *trans*-states into *cis*-states; $\tau_{1/2} = \ln 2/p$, k is Boltzmann constant, h is Plank constant, and T is the temperature. Table 1 shows that $\tau_{1/2}^1 = 3.022$ min for films illuminated for a time less than t_c, while $\tau_{1/2}^1 = 70.692$ min for those films kept under irradiation for time-periods longer than t_c. Equation (2) indicates that $\Delta E_a = 2.012$ eV *at* $0 < t < t_c$ *and* $= 2.093$ eV at $(t_\infty > t > t_c)$, respectively.

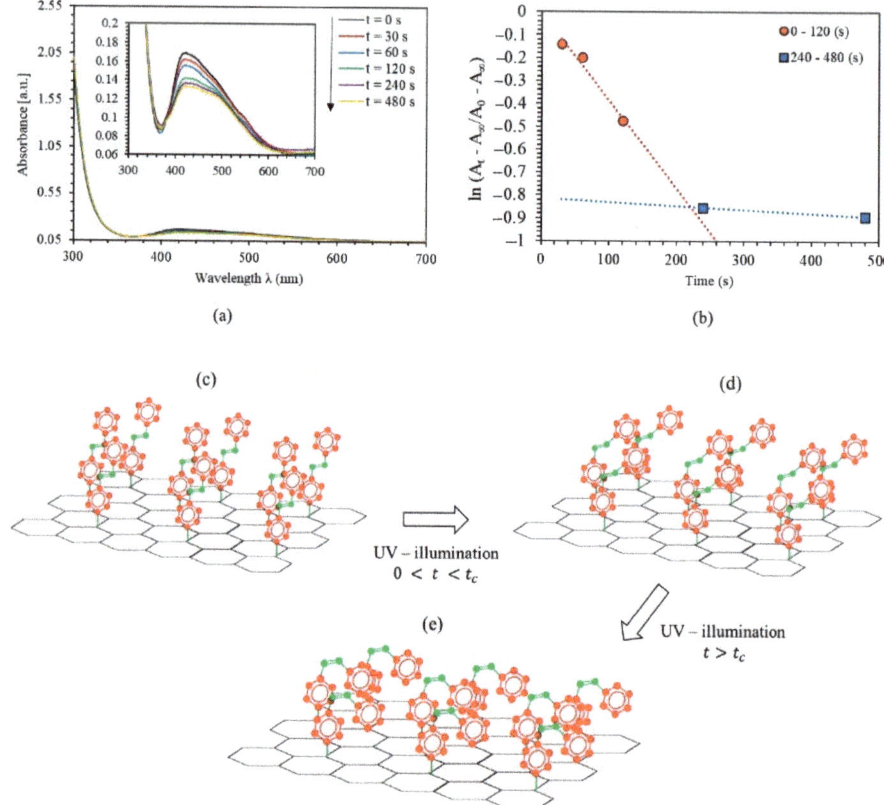

Figure 2. (**a**) Absorbance spectra of PEO-(MR–Graphene) hybrid composite thin film for various UV-illumination times, (**b**) the kinetic constants for *trans* → *cis* photoisomerization of PEO-(MR–Graphene) hybrid composite thin film, (**c**) the chemical structures for the *trans* AZO-Graphene hybrid, and (**d,e**) the *cis* AZO- Graphene hybrid before and after t_c.

Table 1. Kinetic constant (p) and thermal energy barrier (ΔE_a) of PEO-(MR–Graphene) hybrid composite thin film ($t_c = 240$ s).

	$0<t<t_c$			$t_\infty>t>t_c$		
	p_1 (s^{-1})	$\tau^1_{1/2}$ (min)	ΔE_a (eV)	p_2 (s^{-1})	$\tau^2_{1/2}$ (min)	ΔE_a (eV)
AZO–Graphene	3.822×10^{-3}	3.022	2.012	1.634×10^{-4}	70.692	2.093

Beyond t_c, the isomerization energy barrier is higher than that below, and the process proceeds even slower. The calculations agree well with the observed red-shift for the band $n–\pi^*$ in the absorption spectrum beyond t_c where the *cis*-isomer contents have exceeded the *trans*-isomer contents in the PEO-(MR–Graphene) hybrid nanocomposite thin film. Therefore, the separation between the two neighboring *trans*-isomers gets longer, and the intermolecular interactions reduce accordingly. In this scenario, the transformation from the *trans*-to-*cis* photoisomerization is limited to its steric effect, and consequently, the reaction barriers are primarily dominated by the electronic configuration of the $-N = N-$ group [3]. Based on the related findings in the literature [3,33], one may plot the chemical structure of AZO-Graphene used in this work, as shown in Figure 2c. The chemical structure alters as the period of UV-illumination increases leading to enhanced transformation from *trans*-states to *cis*-states. Figure 2d,e shows the *cis*-isomerization of the AZO-Graphene hybrid composite prior to and beyond the t_c reaching the 95.5%.

Another evidence for the kinematic transformation from *trans*-state to *cis*-state for the PEO-(MR–Graphene) hybrid composite thin film is revealed by analyzing the FTIR spectrum. Figure 3 shows the FTIR spectra for the *trans*- and *cis*-states in the spectral range of 500–4000 cm^{-1}. It was observed that the spectral peak for the O–H bond found at 3732 cm^{-1} in the *trans*–state had disappeared in the *cis*–state. This indicated that the free radicals generated by the films' UV irradiation had broken the weak O–H bonds in a photoreaction process. Moreover, the transmittance spectra as a function of frequency for *cis*- and trans-MR showed nearly similar trends with small portions of the bands that were shifted to lower energies. This reflection confirmed that such interactions had a slight influence on N=N bands at frequencies between 1400–1600 cm^{-1}. The evaluated intensity in the infra-red region of N=N band for the *cis*-MR was enormously high compared to that for the *trans*-MR, confirming the enhanced photoisomerization process of the PEO-(MR–Graphene) hybrid nanocomposite thin film.

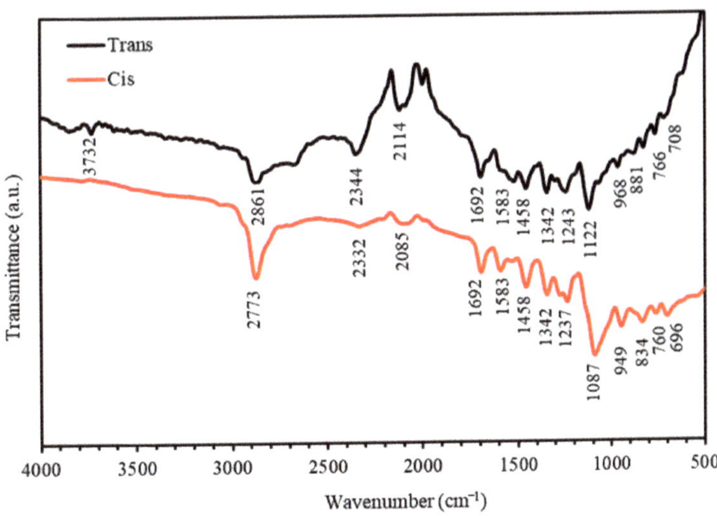

Figure 3. The FTIR spectra of *trans*- and *cis*-states of PEO-(MR–Graphene) hybrid composite thin films.

Furthermore, Figure 2a indicates that each absorbance peak of the PEO-(MR–Graphene) hybrid nanocomposite thin film contains two single absorption peaks (two frequency bands). As the UV-illumination duration increased, the absorption amplitude decreased, indicating that the transformation from *trans*-state to *cis*-state had occurred accordingly. Increasing the duration of the UV-light exposure leads to developing new configured dual-shapes in the absorption peaks shown in Figure 2a, which indicates the development of an impeded dual-frequency bands forming two distinct crests instead of one main absorption peak. Additional evidence for the existence of dual overlapped sub-peaks was revealed from the clear shoulders that appeared mainly at the lower and higher frequency spectrum segments. Namely, at 470 nm and 422.5 nm, respectively.

Figure 4a–f shows the major peaks of PEO-(MR–Graphene) hybrid nanocomposite thin films for all durations of the UV-light exposure in the visible range of the spectrum as fitted to pair Gaussian crests [9]. Figure 4a describes the major peak of PEO-(MR–Graphene) hybrid nanocomposite thin films in the starting *trans*-state fitted to the dual Gaussian crests. The absorption spectra exhibited two bands (high- and low-frequency) with maxima starting at around 417.89 and 471.13 nm and line widths of 53.58 and 120.26 nm, respectively. The absorbance spectrum changed in its configured shape as the UV-illumination exposer time increased, as shown in Figure 4b–f showing the transformation from the *trans*-state to the *cis*-state.

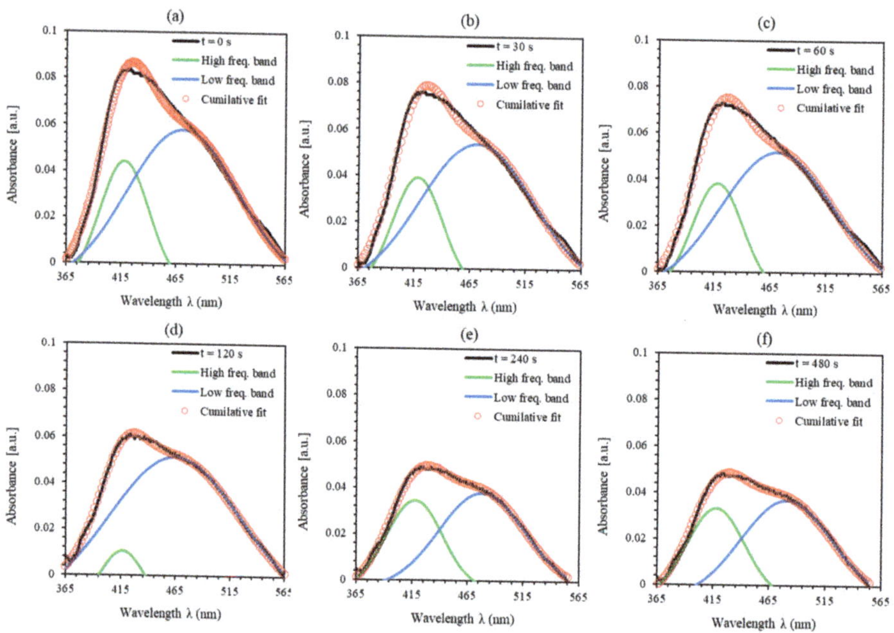

Figure 4. Double peaks fit for the absorbance spectra of PEO-(MR–Graphene) hybrid nanocomposite thin film for various UV-illumination exposure periods (**a**) 0 s, (**b**) 30 s, (**c**) 60 s, (**d**) 120 s, (**e**) 240 s and (**f**) 480 s.

A detailed quantitative analysis for the lower- and higher-frequency bands was performed to obtain a more in-depth insight into the influence of the UV-illumination time. A closer look at Figure 4 shows the specific detailed structure for each peak specifying the amplitude and the area variation underneath the absorbance curve as plotted in Figure 5a,b during the transformation from *trans*- to *cis*-isomerization. Each sub-band's single configuration behavior varies with the UV-light exposure duration as generated through the systematic deviation from the line-width in the related Gaussian function. As can be seen from Figure 5a,b, the area and the amplitude of overall absorbance, low-frequency, high-frequency bands decrease continuously with time till the t_c is reached. However, beyond the t_c, the amplitude and area under the main overall absorbance curve for the low-, high-frequency bands become constant, confirming the achievement of 95.5% proportion of the *cis*-hybrid in AZO-Graphene hybrid nanocomposite. Moreover, the high-frequency sub-band has suffered from a blue-shifting process, which indicates a bathochromic change occurred due to the light absorbed by MR. This is considered as an H-bond donor source or a dissociated-intermediate H-bond donor source [40,41]. Consequently, the overall absorption band is anticipated to occur due to the H-bond's mutual interaction among the PEO and the azo-nitrogen contents through MR molecules. The blue-shifting for both sub-bands (low- and high-frequency bands) exhibit similar trends indicating that H-bonds have a fundamental influence on MR molecules' various sites. Moreover, the high-frequency band demonstrates steady bathochromic shifts during the UV-illumination process, demonstrating two unambiguous symmetries at the H-bonds' backgrounds accompanied by MR transformation. The shift in all bands' amplitude ensures that the time-dependent photoisomerization process is achieved by more than one step [42]. This is because the photoisomerization process is a complex response to the four-level model of *trans*- and *cis*- isomerization as demonstrated by Sekkat Z. et al. [23] and Lee G.T. et al. [43], as shown in Figure 6. The four-levels model describing the *trans*- and the *cis*- isomerization has been investigated thoroughly by Al-Bataineh Q.M. [9]. As is well known, the ground state of AZO-band is a singlet (S0) while the S1 and S2 are first and second singlet excited states, respectively. The S1 state can be generated by either direct excited transition from S0 to S1 or via intersystem crossing between S1 and S2 states (i.e., relaxation of the S2 state down to S1) seen in Figure 6. The states S1 and S2 generated via trans–AZO excitations are different from those of cis–AZO excited states in their energy and conformations.

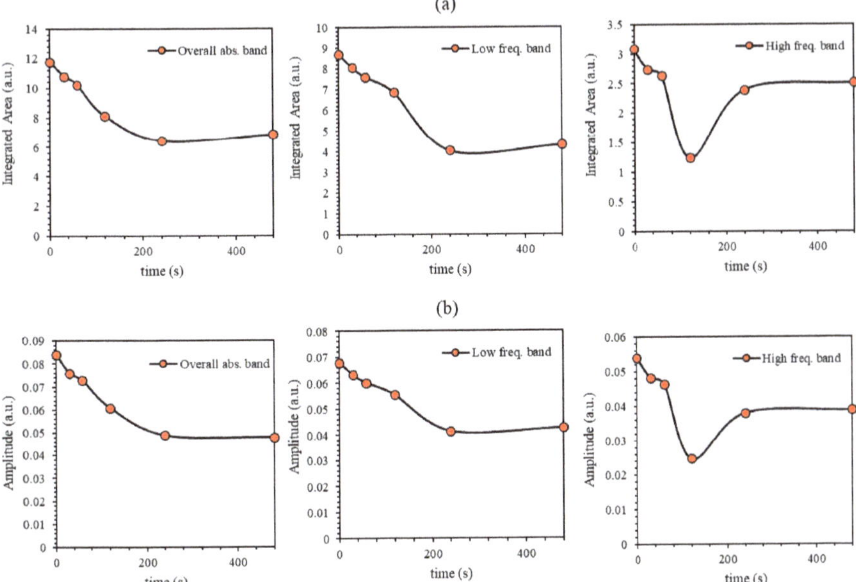

Figure 5. (a) The integrated area and (b) the amplitude-fit for the dual-peaks of the absorbance for the PEO-(MR–Graphene) hybrid nanocomposite thin film for various UV-light exposure periods.

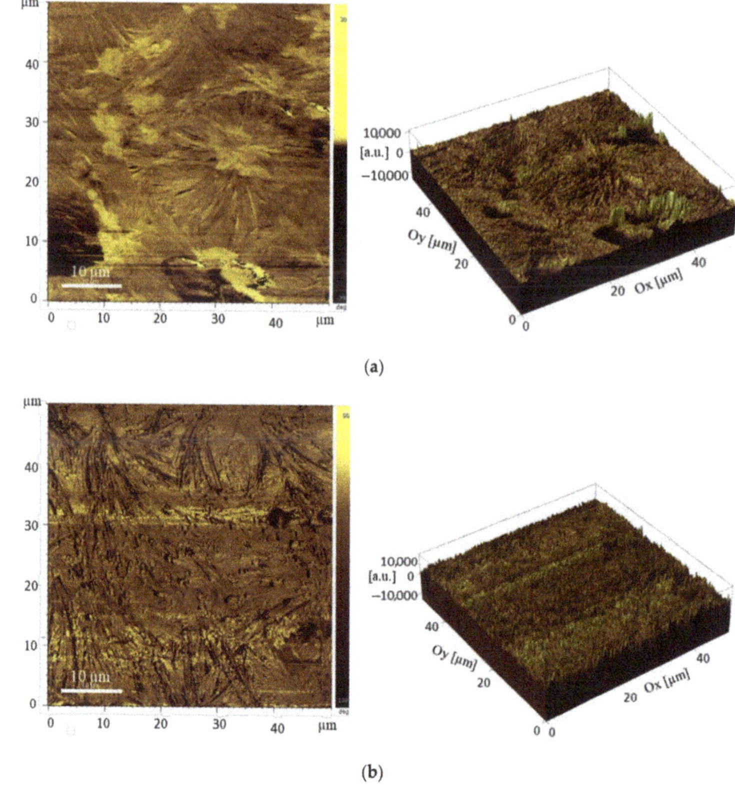

Figure 9. 2D and 3D AFM images of PEO-(MR–Graphene) hybrid nanocomposite thin films: (**a**) without curing and (**b**) with curing.

4. Conclusions

The fundamental mechanisms of the kinetics of photoisomerization of the PEO-(MR-Graphene) hybrid nanocomposite thin films are explored, evaluated, and investigated thoroughly. Deliberately, we explore and provide a new insight into the photoisomerization kinetics and time evolution for a hybrid thin-film based on Azo dye methyl red (MR) incorporated with graphene hosted in polyethylene oxide. The kinetics of photoisomerization and time evolution of the hybrid thin film was examined via the UV-Vis and the FTIR spectroscopic techniques and by using specific analytical models. The existence of the Azo Dye MR in the amalgamated composites is crucial for the effectual acts of the *trans* ↔ *cis* cyclic isomerization via UV-illumination ↔ Visible light-relaxation. Moreover, the UV–Vis and FTIR investigations confirm the hysteresis cycles of *trans-cis*-states. In conclusion, the PEO-(MR-Graphene) hybrid nanocomposite thin films are proven to be potential candidates for many applications. These applications include photochromic molecular switches, light-gated transistors, in addition to molecular solar thermal energy storage media.

Author Contributions: Conceptualization, Q.M.A.-B., A.A.A. and A.T.; Data curation, Q.M.A.-B. and A.M.A.; Formal analysis, Q.M.A.-B., A.A.A., A.M.A. and A.T.; Funding acquisition, A.A.A., A.M.A. and A.T.; Investigation, Q.M.A.-B. and A.T.; Methodology, Q.M.A.-B., A.A.A., A.M.A. and A.T.; Project administration, A.A.A., A.M.A. and A.T.; Resources, A.A.A.; Supervision, Q.M.A.-B.; Validation, A.T.; Visualization, Q.M.A.-B. and A.A.A.; Writing–original draft, Q.M.A.-B.; Writing–review & editing, A.A.A., A.M.A. and A.T. All authors have read and agreed to the published version of the manuscript.

Funding: This research received no external funding.

Acknowledgments: The authors would like to acknowledge Jordan University of Science and Technology (JUST) in Jordan. Our thanks are extended to Mohammad-Ali H. Al-Akhras for helping our members to use his biomedical laboratories.

Conflicts of Interest: The authors declare no conflict of interest.

Data Availability: All the data provided in this manuscript are either collected from out apparatus or derived from the raw data collected. They are available at any time for readers.

References

1. Schneider, V.; Strunskus, T.; Elbahri, M.; Faupel, F.; Elbahri, M. Light-induced conductance switching in azobenzene based near-percolated single wall carbon nanotube/polymer composites. *Carbon* **2015**, *90*, 94–101. [CrossRef]
2. Kim, M.; Safron, N.S.; Huang, C.; Arnold, M.S.; Gopalan, P. Light-Driven Reversible Modulation of Doping in Graphene. *Nano Lett.* **2012**, *12*, 182–187. [CrossRef]
3. Pang, W.; Xue, J.; Pang, H. A High Energy Density Azobenzene/Graphene Oxide Hybrid with Weak Nonbonding Interactions for Solar Thermal Storage. *Sci. Rep.* **2019**, *9*, 5224. [CrossRef]
4. Schuschke, C.; Hohner, C.; Jevric, M.; Petersen, A.U.; Wang, Z.; Schwarz, M.; Kettner, M.; Waidhas, F.; Fromm, L.; Sumby, C.J.; et al. Solar energy storage at an atomically defined organic-oxide hybrid interface. *Nat. Commun.* **2019**, *10*, 2384. [CrossRef]
5. Mansø, M.; Petersen, A.U.; Wang, Z.; Erhart, P.; Nielsen, M.B.; Moth-Poulsen, K. Molecular solar thermal energy storage in photoswitch oligomers increases energy densities and storage times. *Nat. Commun.* **2018**, *9*, 1945. [CrossRef]
6. Wang, Z.; Udmark, J.; Börjesson, K.; Rodrigues, R.; Roffey, A.; Abrahamsson, M.; Nielsen, M.B.; Moth-Poulsen, K. Evaluating Dihydroazulene/Vinylheptafulvene Photoswitches for Solar Energy Storage Applications. *ChemSusChem* **2017**, *10*, 3049–3055. [CrossRef] [PubMed]
7. Alsaad, A.; Al-Bataineh, Q.M.; Telfah, M.; Ahmad, A.A.; AlBataineh, Z.; Telfah, A. Optical properties and photo-isomerization processes of PMMA–BDK–MR nanocomposite thin films doped by silica nanoparticles. *Polym. Bull.* **2020**, 1–17. [CrossRef]
8. Blinov, L.M.; Kawai, T.; Kozlovsky, M.V.; Kawata, Y.; Ichimura, K.; Seki, T.; Tripathy, S.; Li, L.; Oliveira, O., Jr.; Irie, M. *Photoreactive Organic Thin Films*; Elsevier: Amsterdam, The Netherlands, 2002.
9. AlBataineh, Z.; Ahmad, A.A.; Alsaad, A.; Qattan, I.A.; Bani-Salameh, A.A.; Telfah, A. Kinematics of Photoisomerization Processes of PMMA-BDK-MR Polymer Composite Thin Films. *Polymers* **2020**, *12*, 1275. [CrossRef]
10. Ahmad, A.A.; Alsaad, A.M.; Al-Bataineh, Q.M.; Al-Akhras, M.-A.H.; AlBataineh, Z.; Alizzy, K.A.; Daoud, N.S. Synthesis and characterization of ZnO NPs-doped PMMA-BDK-MR polymer-coated thin films with UV curing for optical data storage applications. *Polym. Bull.* **2020**, 1–23. [CrossRef]
11. Morgenstern, K. Isomerization reactions on single adsorbed molecules. *Acc. Chem. Res.* **2009**, *42*, 213–223. [CrossRef]
12. Henzl, J.; Mehlhorn, M.; Gawronski, H.; Rieder, K.H.; Morgenstern, K. Reversible cis–trans isomerization of a single azobenzene molecule. *Angew. Chem. Int. Ed.* **2006**, *45*, 603–606. [CrossRef] [PubMed]
13. Choi, B.-Y.; Kahng, S.-J.; Kim, S.; Kim, H.; Kim, H.W.; Song, Y.J.; Ihm, J.; Kuk, Y. Conformational Molecular Switch of the Azobenzene Molecule: A Scanning Tunneling Microscopy Study. *Phys. Rev. Lett.* **2006**, *96*, 156106. [CrossRef] [PubMed]
14. Georgiou, D.; Melidis, P.; Aivasidis, A.; Gimouhopoulos, K. Degradation of azo-reactive dyes by ultraviolet radiation in the presence of hydrogen peroxide. *Dye. Pigment.* **2002**, *52*, 69–78. [CrossRef]
15. Ashter, S.A. *Introduction to Bioplastics Engineering*; Elsevier: Amsterdam, The Netherlands, 2016.
16. Ashter, S.A. 7-processing biodegradable polymers. In *Introduction to Bioplastics Engineering*; William Andrew Publishing: Oxford, UK, 2016; pp. 179–209.
17. Izdebska, J. Aging and Degradation of Printed Materials. *Print. Polym.* **2016**, 353–370. [CrossRef]
18. Ebrahimi, H.R.; Modrek, M. Photocatalytic Decomposition of Methyl Red Dye by Using Nanosized Zinc Oxide Deposited on Glass Beads in Various pH and Various Atmosphere. *J. Chem.* **2013**, *2013*, 151034. [CrossRef]

19. Yaqoob, A.A.; Serrà, A.; Ibrahim, M.N.M. Advances and Challenges in Developing Efficient Graphene Oxide-Based ZnO Photocatalysts for Dye Photo-Oxidation. *Nanomaterials* **2020**, *10*, 932. [CrossRef]

20. Zrnčić, M.; Ljubas, D.; Ćurković, L.; Škorić, I.; Babić, S. Kinetics and degradation pathways of photolytic and photocatalytic oxidation of the anthelmintic drug praziquantel. *J. Hazard. Mater.* **2017**, *323*, 500–512.

21. Anwer, H.; Mahmood, A.; Lee, J.; Kim, K.-H.; Park, J.-W.; Yip, A.C.K. Photocatalysts for degradation of dyes in industrial effluents: Opportunities and challenges. *Nano Res.* **2019**, *12*, 955–972. [CrossRef]

22. Shi, Y.; Steier, W.H.; Yu, L.; Chen, M.; Dalton, L.R. Large stable photoinduced refractive index change in a nonlinear optical polyester polymer with disperse red side groups. *Appl. Phys. Lett.* **1991**, *58*, 1131–1133. [CrossRef]

23. Sekkat, Z.; Morichère, D.; Dumont, M.; Loucif-Saïbi, R.; Delaire, J.A. Photoisomerization of azobenzene derivatives in polymeric thin films. *J. Appl. Phys.* **1992**, *71*, 1543–1545. [CrossRef]

24. Zhang, X.; Feng, Y.; Huang, D.; Li, Y.; Feng, W. Investigation of optical modulated conductance effects based on a graphene oxide–azobenzene hybrid. *Carbon* **2010**, *48*, 3236–3241. [CrossRef]

25. Coelho, P.J.; Sousa, C.M.; Castro, M.C.R.; Fonseca, A.M.C.; Raposo, M.M.M. Fast thermal cis–trans isomerization of heterocyclic azo dyes in PMMA polymers. *Opt. Mater.* **2013**, *35*, 1167–1172. [CrossRef]

26. Feringa, B.; Jager, W.F.; De Lange, B. Organic materials for reversible optical data storage. *Tetrahedron* **1993**, *49*, 8267–8310. [CrossRef]

27. Karmakar, A.; Ghosh, A. Charge carrier dynamics and relaxation in (polyethylene oxide-lithium-salt)-based polymer electrolyte containing 1-butyl-1-methylpyrrolidinium bis(trifluoromethylsulfonyl)imide as ionic liquid. *Phys. Rev. E* **2011**, *84*, 051802. [CrossRef] [PubMed]

28. Morsi, M.; Rajeh, A.; Al-Muntaser, A. Reinforcement of the optical, thermal and electrical properties of PEO based on MWCNTs/Au hybrid fillers: Nanodielectric materials for organoelectronic devices. *Compos. Part B Eng.* **2019**, *173*, 106957. [CrossRef]

29. Abutalib, M.; Rajeh, A. Influence of ZnO/Ag nanoparticles doping on the structural, thermal, optical and electrical properties of PAM/PEO composite. *Phys. B Condens. Matter* **2020**, *578*, 411796. [CrossRef]

30. Hadi, A.; Hashim, A.; Al-Khafaji, Y. Structural, Optical and Electrical Properties of PVA/PEO/SnO2 New Nanocomposites for Flexible Devices. *Trans. Electr. Electron. Mater.* **2020**, *21*, 283–292. [CrossRef]

31. Li, Y.; Chen, Z. XH/π (X = C, Si) interactions in graphene and silicene: Weak in strength, strong in tuning band structures. *J. Phys. Chem. Lett.* **2013**, *4*, 269–275. [CrossRef]

32. Dutta, P.; Nandi, D.; Datta, S.; Chakraborty, S.; Das, N.; Chatterjee, S.; Ghosh, U.C.; Halder, A. Excitation wavelength dependent UV fluorescence of dispersed modified graphene oxide: Effect of pH. *J. Lumin.* **2015**, *168*, 269–275. [CrossRef]

33. Wang, D.; Ye, G.; Wang, X.; Wang, X. Graphene Functionalized with Azo Polymer Brushes: Surface-Initiated Polymerization and Photoresponsive Properties. *Adv. Mater.* **2011**, *23*, 1122–1125. [CrossRef]

34. Saladaga, I.A.; Janthasit, S.; de Luna, M.D.G.; Grisdanurak, N.; Tantrawong, S.; Paoprasert, P. Poly (oligoethylene Glycol Methacrylate): A Promising Electrolyte Polymer. *Chiang Mai J. Sci.* **2015**, *42*, 868–876.

35. Saeed, K.; Ishaq, M.; Ilyas, M. Preparation, morphology, and thermomechanical properties of coal ash/polyethylene oxide composites. *Turk. J. Chem.* **2011**, *35*, 237–243.

36. Kucharski, T.J.; Tian, Y.; Akbulatov, S.; Boulatov, R. Chemical solutions for the closed-cycle storage of solar energy. *Energy Environ. Sci.* **2011**, *4*, 4449. [CrossRef]

37. Feng, Y.; Liu, H.; Luo, W.; Liu, E.; Zhao, N.; Yoshino, K.; Feng, W. Covalent functionalization of graphene by azobenzene with molecular hydrogen bonds for long-term solar thermal storage. *Sci. Rep.* **2013**, *3*, 3260. [CrossRef]

38. Shin, K.-H.; Shin, E.J. Photoresponsive Azobenzene-modified Gold Nanoparticle. *Bull. Korean Chem. Soc.* **2008**, *29*, 1259–1262.

39. Yang, Y.; Hughes, R.P.; Aprahamian, I. Visible Light Switching of a BF2-Coordinated Azo Compound. *J. Am. Chem. Soc.* **2012**, *134*, 15221–15224. [CrossRef]

40. Reichardt, C. Solvatochromic Dyes as Solvent Polarity Indicators. *Chem. Rev.* **1994**, *94*, 2319–2358. [CrossRef]

41. Ozen, A.S.; Doruker, P.; Aviyente, V. Effect of Cooperative Hydrogen Bonding in Azo–Hydrazone Tautomerism of Azo Dyes. *J. Phys. Chem. A* **2007**, *111*, 13506–13514. [CrossRef]

42. Ruzza, P.; Hussain, R.; Biondi, B.; Calderan, A.; Tessari, I.; Bubacco, L.; Siligardi, G. Effects of Trehalose on Thermodynamic Properties of Alpha-synuclein Revealed through Synchrotron Radiation Circular Dichroism. *Biomolecules* **2015**, *5*, 724–734. [CrossRef]
43. Lee, G.J.; Kim, N.; Lee, M. Photophysical properties and photoisomerization processes of Methyl Red embedded in rigid polymer. *Appl. Opt.* **1995**, *34*, 138–143. [CrossRef]
44. Airinei, A.; Buruiana, E.C. Photoisomerization of AZO Aromatic Chromophobes in Polyvinyl Chloride. *J. Macromol. Sci. Part A* **1994**, *31*, 1233–1239. [CrossRef]

Publisher's Note: MDPI stays neutral with regard to jurisdictional claims in published maps and institutional affiliations.

 *polymers*

Article

Applying Membrane Distillation for the Recovery of Nitrate from Saline Water Using PVDF Membranes Modified as Superhydrophobic Membranes

Fatemeh Ebrahimi [1,2], **Yasin Orooji** [1,*] and **Amir Razmjou** [3,4,*]

1   College of Materials Science and Engineering, Nanjing Forestry University, Nanjing 210037, China; rto.est@gmail.com

2   Department of Nanotechnology, Faculty of Advanced Sciences and Technologies, University of Isfahan, Isfahan 73441-81746, Iran

3   Department of Biotechnology, Faculty of Advanced Sciences and Technologies, University of Isfahan, Isfahan 73441-81746, Iran

4   UNESCO Centre for Membrane Science and Technology, School of Chemical Science and Engineering, University of New South Wales, Sydney 2052, Australia

*   Correspondence: yasin@njfu.edu.cn (Y.O.); amirr@unsw.edu.au (A.R.)

Received: 23 October 2020; Accepted: 19 November 2020; Published: 24 November 2020

Abstract: In this study, a flat sheet direct contact membrane distillation (DCMD) module was designed to eliminate nitrate from water. A polyvinylidene fluoride (PVDF) membrane was used in a DCMD process at an ambient pressure and at a temperature lower than the boiling point of water. The electrical conductivity of the feed containing nitrate increased, while the electrical conductivity of the permeate remained constant during the entire process. The results indicated that the nitrate ions failed to pass through the membrane and their concentration in the feed increased as pure water passed through the membrane. Consequently, the membrane was modified using TiO_2 nanoparticles to make a hierarchical surface with multi-layer roughness on the micro/nanoscales. Furthermore, 1H,1H,2H,2H-Perfluorododecyltrichlorosilane (FTCS) was added to the modified surface to change its hydrophobic properties into superhydrophobic properties and to improve its performance. The results for both membranes were compared and reported on a pilot scale using MATLAB. In the experimental scale (a membrane surface area of 0.0014 m², temperature of 77 °C, nitrate concentration of 0.9 g/Kg, and flow rate of 0.0032 Kg/s), the flux was 2.3 $Kgm^{-2}h^{-1}$. The simulation results of MATLAB using these data showed that for the removal of nitrate (with a concentration of 35 g/Kg) from the intake feed with a flow rate of 1 Kg/s and flux of 0.96 $Kgm^{-2}h^{-1}$, a membrane surface area of 0.5 m² was needed.

Keywords: nanocomposite membrane; membrane distillation; nitrate; sustainable development; simulation

---

1. Introduction

The vast variety of nanotechnology applications is triggering a breakthrough across many fields [1–7]. Water and water resources are among the fundamental pillars of sustainable development in any country [8–10]. Thus, they should be preserved and managed in such a way that the needs of future generations for the supply of this essential material are taken into account [11]. The shortage of freshwater supplies on the earth and climate change leading to recent droughts in some countries have turned water scarcity into a global and environmental concern [12]. This shortage entails the necessity of paying more attention to preserving the existing resources and recycling wastewaters [12]. In this regard, the quantitative and qualitative preservations of water resources especially in dry and waterless areas as well as soil preservation are of great importance in the survival of the earth and the living organisms on it [13–15]. Furthermore, the infiltration of different types of pollutants as well as urban, industrial, and agricultural wastewaters into the groundwater have had long-term adverse effects on the quality of water resources and people's health [16–18]. Chemical materials play a significant role in the contamination of the groundwater and wastewater. A considerable number of these pollutants (especially nitrate) originate from different types of fertilizers used in agricultural activities [19,20].

The consumption of water polluted with nitrate is harmful to human health (especially to children) and sometimes causes severe harms and even death. Gastrointestinal bacteria in the stomach have the ability to transform nitrate to nitrite. A high amount of nitrite in the body can lead to methemoglobin disease, infant death, and abortion. Nitrite can also be combined with amines or amides in the body and form nitrosamine, a well-known carcinogen [21,22]. Hence, nitrate removal from drinking water is of great importance from environmental and health perspectives. Although water pollution with nitrate has several reasons, the use of nitrate fertilizers and pesticides in agriculture is considered as one of the most important reasons [23–25]. According to the World Health Organization (WHO), the permitted nitrate level in drinking water is 50 mg/L (in nitrate). Furthermore, the United States Environmental Protection Agency has announced the maximum allowed level as 10 mg/L in nitrogen equivalent to 44.82 mg/L in nitrate [26]. Nitrate can also impose serious environment issues. Over-supply of nitrate-based fertilizers results in a contamination of soil which makes the farmlands unsuitable for future farming. In addition, the high concentration of nitrate in water resources causes the reduction of its oxygen level damaging aquatic life.

Membrane distillation is a relatively new method for purifying solutions and treating water. This method which is a combination of distillation and membrane separation methods not only possesses the advantages of both of these methods but also lacks their disadvantages to a large extent. The basis of separation in both distillation and membrane distillation processes is the liquid–vapor equilibrium. Moreover, they both change their phases by receiving the latent evaporation heat. In conventional distillation, it is necessary to heat the feed to the boiling point, while the membrane distillation process is carried out at a temperature below the boiling point; therefore, energy is saved. Membrane distillation is a separation method based on using a membrane and has solved many problems of other membrane separation methods, such as polarization and increased concentration, high energy requirements, and the need for multiple purification steps.

The driving force of membrane distillation is the temperature difference between the two sides of the membrane which leads to a difference in the vapor pressure on its two sides and the passage of water vapor through it. Water cannot move to the other side because the membrane is hydrophobic or superhydrophobic. Additionally, the surface tension force of water and the repulsive force between the water droplets and the hydrophobic membrane used in this process prevent water droplets from passing through the membrane pores. However, water vapor can easily pass through them. In other words, instead of using the distillation column, the pore volume of the porous membrane is used which reduces the costs and saves space and equipment. Since this process is based on the passage of vapor, non-evaporable ions, colloids, and macromolecules cannot pass through the membrane and return to the solution. Additionally, other materials with a boiling point higher than the feed temperature cannot pass through the membrane. As a result, the permeate is highly pure. Thus, in this process, water is separated without being wasted and is independent of the feed concentration [27]. The hot feed and the cold pure water flow on the two sides of the membrane in opposite directions. Another advantage of the counter-current module is that the chance of membrane fouling by solid materials in this method is less than that of other methods. As a result, the membrane becomes more durable [28]. Due to the rotational water flow in this process, the residual thermal energy of the feed is not wasted and returns to the system.

The main advantage of this process is that it is done at atmospheric pressure at a much lower temperature than the boiling point of water. This advantage has led to widespread applications for this method. Meanwhile, to supply this low temperature, renewable energy sources including geothermal and solar energies may be employed at household or industrial scales. Direct Contact Membrane

Distillation (DCMD), Sweeping Gas Membrane Distillation (SGMD), Air Gap Membrane Distillation (AGMD), and Vacuum Membrane Distillation (VMD) are four MD configurations.

Razmjou et al. (2012) reported that they managed to make a PVDF membrane superhydrophobic in order to apply it in DCMD membrane distillation [29]. They found in their study that pore wetting was a major problem in membrane distillation. When the pores get wet, the feed partially or entirely passes through the pores, contaminates the permeate, and blocks the pores [30]. Hence, the cross-membrane flux and the membrane performance are reduced. To solve this problem, they engineered the surface of the hydrophobic membrane and converted it into a superhydrophobic one. This modification significantly reduced the direct contact surface between the membrane and the feed which reduced the wettability of the pores, their contamination, heat loss near the membrane, and also the driving force required by the process. Boubakri et al. (2013) removed nitrate from water with a purity of 99.90% by direct membrane distillation and using polypropylene (PP) and polyvinylidene fluoride (PVDF) hydrophobic membranes [27]. Their results showed that the limitations of the methods mentioned above including the relationship between nitrate removal and anions with nitrate and nitrate concentration did not affect the results of the process.

Zhang et al. (2013) made a superhydrophobic PVDF membrane to apply it in DCMD [31]. They developed a superhydrophobic composite membrane by coating a mixture of PolyDiMethylSiloxane (PDMS) and SiO2 hydrophobic nanoparticles on a PVDF membrane so that the contact angle changed from 107 to 156 degrees. Although the cross-membrane water flux of the modified membrane was reduced compared to that of virgin membrane, the permeate was purified of sodium chloride with a purification of 99.99%. In addition, the modified membrane showed significant anti-fouling properties. Dong et al. (2016) simulated the removal of NaCl from water by membrane distillation technology [18] on the industrial scale.

The aim of this research was to optimize the hydrophobic properties of the membrane in two stages: first, changing the topography of the surface to enhance roughness by coating TiO_2 nanoparticles on the surface of the membrane; second, changing the chemical properties of the surface to repel water more using the functional agent FTCS. The modified membrane was used in a DCMD membrane module in the experimental scale. Then the results of this optimization were evaluated by using Dong's MATLAB codes for NaCl [18]. Because water polluted with NaCl and water polluted with nitrate are similar in terms of physical properties, such as density and the number of ions, these codes were used for the feed containing nitrate. Then, the obtained experimental results were given to MATLAB software which changed them into the semi-industrial scale.

2. Materials and Methods

2.1. Materials

In this study, the following materials (provided from Sigma-Aldrich) were used: polyvinylidene fluoride (PVDF) flat-sheet membrane HVHP (Millipore, nominal pore size: 0.45 µm, porosity: 75%), titanium (IV) isopropoxide (TTIP) (98%) as TiO_2 precursor, 2,4-pentanedione, acetylacetone, Milli-Q Water, perchloric acid (70%), 1H,1H,2H,2H-perfluorododecyltrichlorosilane (FTCS), potassium nitrate, potassium iodide, absolute ethanol, and toluene.

2.2. Membrane Surface Modification

In order to synthesize titanium oxide nanoparticles (Figure 1), ethanol, 2,4-pentanedione, perchloric acid 70%, titanium tetraisopropoxide, and Milli-Q water were mixed, at room temperature to form a stable sol of the nanoparticles. The sol was stirred for an hour. The molar ratios of each component in the resulting sol were TTIP: Pluronic F127: 2,4-pentanedione: $HClO_4$:H_2O: Ethanol = 1:0.004:0.5:0.5:0.45:4.76. PEG (1000 g/mol) is also substituted with the Pluronic F127 to see the effect of hydrophilic templating agent on the coating films [32].

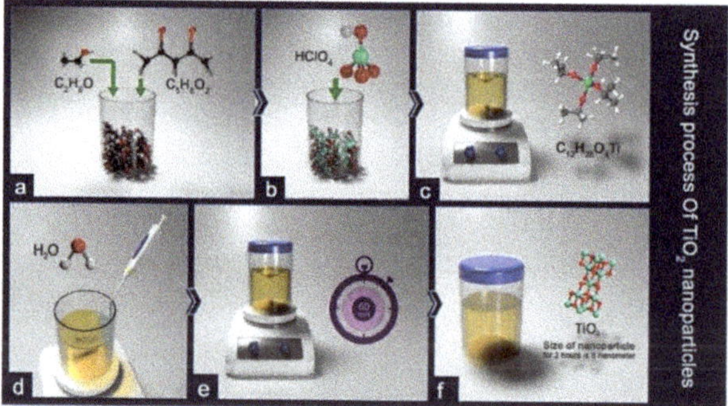

Figure 1. The synthesis of TiO$_2$ nanoparticles using the sol-gel method: (**a**) mixing ethanol and 2,4-pentanedione; (**b**) adding perchloric acid 70%; (**c**) stirring the solution and adding titanium tetraisopropoxide; (**d**) adding Milli-Q water dropwise; (**e**) stirring the solution for an hour at room temperature; (**f**) the TiO$_2$ sol gel solution.

The operation was carried out by the dip-coating process. The stages of this process are explained below (Figure 2). The dip coater was used to coat TiO$_2$ nanoparticles on the membrane by dipping the membrane in the sol-gel solution at a speed of 50 mm/s and holding it in the solution for 8 s. Then, the membrane was removed at the same speed. Afterward, the sample was dried in an oven at 120 °C for one hour. Some nanoparticles do not create a strong chemical bond during the coating process or merely create a weak physical bond with the surface. The hydrothermal process is performed to increase the energy of these particles to make a strong bond with the surface after coating the nanoparticles on the membrane surface. The membrane coated with a sol-gel solution was placed in a sealed container overflowed with Milli-Q water and was placed in an autoclave at 90 °C for 2 h. After coating and heat treatment, the sample was radiated with ultraviolet light (UV) for 6 h. The residual organic sediments were decomposed by this radiation. Finally, the membrane was called 'TiO$_2$-PVDF'.

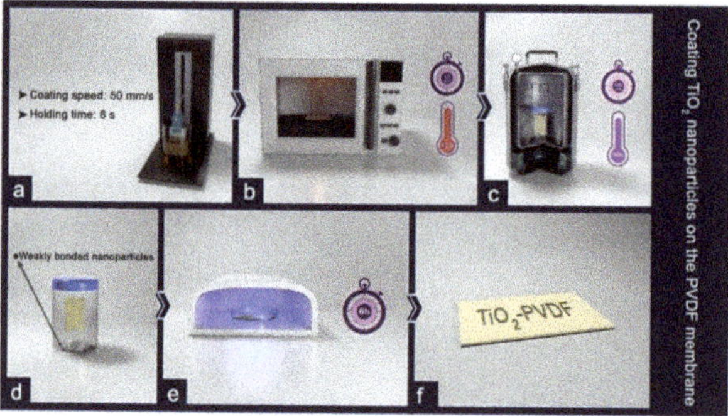

Figure 2. Coating TiO$_2$ nanoparticles on the PVDF membrane: (**a**) dip-coating the membrane in the TiO$_2$ sol-gel solution (coating speed: 50 mm/s; holding time: 8 s); (**b**) drying the membrane at 120 °C for an hour; (**c**) treating the membrane using the hydrothermal method at a low temperature (90 °C) for two hours; (**d**) the separation of the weakly bonded nanoparticles after heat treatment; (**e**) exposing the membrane to UV light for six hours; (**f**) the TiO$_2$–PVDF membrane.

2.3. Fluorination of the Surface of the TiO$_2$–PVDF Membrane

First toluene was purged with nitrogen for one hour to remove its oxygen content as much as possible. Then, to reduce its temperature, toluene was refrigerated for 2 h. FTCS powder was then added to toluene within a controlled temperature in the range of 0 to −5 °C. Then while the solution was stirred it was treated with an ice bath. The FTCS solution was coated on the membrane surface of the TiO$_2$–PVDF nanocomposite by the self-assembly method. In this method, after preparing the FTCS solution, the samples coated with titanium nanoparticles were placed in the solution so that the FTCS molecules were bonded with the TiO$_2$–PVDF membrane by the self-assembly method (Figure 3). Hereafter, the membrane is called the 'FTCS–TiO$_2$–PVDF' membrane.

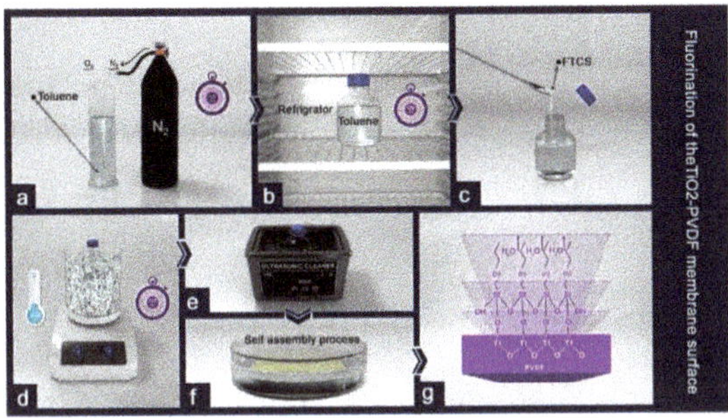

Figure 3. The preparation process of the FTCS–TiO$_2$–PVDF membrane: (**a**) purifying toluene using nitrogen; (**b**) reducing the temperature of toluene in a refrigerator; (**c**) mixing toluene with FTCS; (**d**) stirring the mix at a low temperature (−5 °C–0 °C); (**e**) using ultrasonication and ice bath to dissolve the mix completely; (**f**) the self-assembly process of the FTCS solution on the TiO2–PVDF membrane; (**g**) the FTCS–TiO$_2$–PVDF membrane.

2.4. The Performance of the Membranes with Synthetic Feed Waters

After ensuring the good performance of the module by conducting the hot/cold water test, the main test was conducted at the atmospheric pressure of one for 6 h in two phases with direct contact membrane distillation setup (Figure 4 and Figure S1). The first phase included operations using an unmodified membrane and the second phase involved operations with a modified membrane. The temperature difference of the feed and the permeate was kept constant during the process (about 50 degrees on average). First, water containing nitrate was synthesized by adding 5 wt % potassium nitrate.

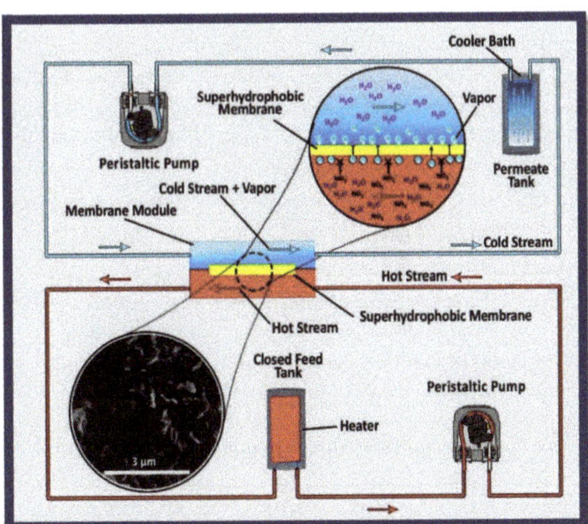

Figure 4. Lab-scale schematic diagram of direct contact membrane distillation setup.

Parallel counter-currents (the hot-water feed containing nitrate ions and the cold water containing pure permeate) were formed using two peristaltic pumps with a constant speed of 37 rpm. The feed container was placed in a hot water tank whose temperature was controlled. The permeate container was placed in a tank containing a mixture of water and ice and its temperature was controlled. The temperatures of the feed and permeate were controlled by isolating the system. The water vapor passed through the membrane from the feed side to the permeate side. In the cold side, vapor lost its energy and became water. During the test, the weight and volume of the feed gradually decreased, while the volume and weight of the permeate increased. The amount of ion in the feed and permeate containers was measured by an electrical conductivity meter every hour. The weights of the feed and permeate were also measured by a scale every hour.

2.5. Pilot Process Simulation

Dong et al. [33] created a mathematical approach by pairing tanks-in-series and a black box to investigate all the main factors of the DCMD process versus the length of the membrane. The designed simulator was used to predict and evaluate the performance of the flat-sheet DCMD at an industrial scale, as well as the impact of the physical characteristics of the membrane, module dimensions, and the implementation of conditions on the performance of the large-scale DCMD module. This research was conducted to obtain the design considerations for the production of pure water. In their study, the software required operational and laboratory-scale data to produce the desired outputs in large-scale countercurrents (Tables S1 and S2 and Figure S2). In the current research, the experimental conditions of Dong et al.'s study were reconstructed. However, nitrate was used instead of NaCl. Due to the similarities between the physical properties of nitrate and NaCl, Dong's MATLAB codes were used to convert experimental results to large-scale results.

3. Results and Discussion

3.1. Surface Morphology

The morphology of membranes was characterized by FE-SEM (QUANTA FEG 450, Hillsboro, OR, USA) operating at 15 kV. The comparison of several scanning electron microscopy (SEM) images of the FTCS–TiO2–PVDF membrane and the PVDF membrane (Figure 5a,b) shows that a multi-level hierarchical roughness was created on the surface of the membrane at the end of the process which prevented water droplets from penetrating the surface. Furthermore, while the contact angle increased substantially, the average pore size did not change significantly (Figure 5c–f). It was expected that the performance of the modified membrane was better than that of the unmodified membrane. Furthermore, the microscopic changes on the surface of the membrane were investigated in different stages.

As can be seen in (Figure 5g,h), coating TiO_2 nanoparticles did not significantly change the porosity, pore size, and surface structure of the membrane. These nanoparticles are important in the creation of superhydrophobic properties in two regards: first, they increase the surface roughness; second, they are an anchor for linking the FTCS molecules on the surface.

The microscopic images of coating FTCS on the membrane (Figure 5i,j) show that the self-assembly method was able to create a more uniform coating on the surface than previously reported methods. As can be seen in this figure, despite the penetration of FTCS into the pores, no significant surface structure changes occurred, and the size of the pores did not change. It should also be noted that preserving the size and structure of the pores is crucial in the performance of DCMD. FTCS was used to reduce surface energy chemically. However, as can be observed in (Figure 5i,j) FTCS also increased hydrophobicity by creating roughness. FTCS crystals increased the surface roughness (Figure 5k,i). Furthermore, the images show that the sizes of the pores did not change substantially (Figure 5i–q) SEM images of different samples of the FTCS–PVDF membrane in different magnifications (Figure 5m–q).

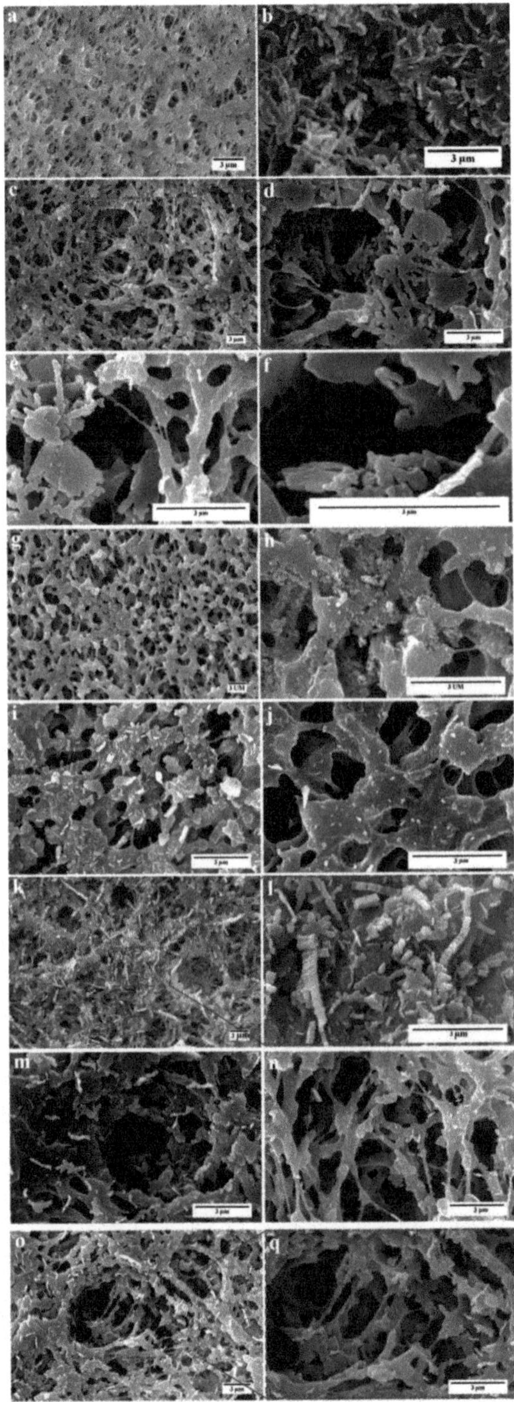

Figure 5. The FESEM images of (**a**) the virgin PVDF membrane; (**b–f**) the FTCS–TiO2–PVDF membrane at different magnifications; (**g,h**) the TiO2–PVDF membrane in different magnifications; (**i–p**) the FTCS–PVDF membrane in different magnifications.

In a study carried out by Razmjou et al. [29], the increase of the contact angle in the stage of coating FTCS on PVDF was reported as 146° ± 5° (the contact angle for the virgin membrane was 125° ± 1°) while in the current work which was done by the self-assembly method, the contact angle was 142° ± 20° (the contact angle for the virgin membrane was 89° ± 8°). The contact angles formed by water droplet on the membranes surface were measured by the sessile drop technique (CA-500A instrument, Yasin Pajooh Co. Ltd., Isfahan, Iran).

3.2. Investigating the Durability and Photocatalytic Activity of TiO$_2$ Nanoparticles Coated on the Surface

The KI test was employed to ascertain the durability of TiO$_2$ nanoparticles on the surface of the PVDF membrane (Figure 6). If the coated membrane has a photocatalytic activity, it can oxidize ion (I$^-$) according to Equation (1) and convert it to (I$_2$) and change the color of the colorless solution KI to yellow color.

$$I^- + h^+ \rightarrow \tfrac{1}{2} I_2$$
$$I_2 + I^- \rightarrow I_3^-$$

(1)

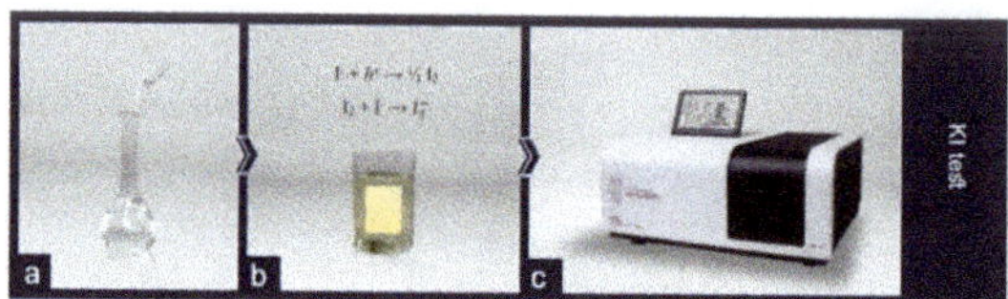

Figure 6. Investigating the durability of TiO$_2$ nanoparticles on the surface of the PVDF membrane using the KI test: (**a**) preparing the KI solution; (**b**) the change in the color of the colorless solution by entering the TiO$_2$–PVDF membrane; (**c**) the UV-Vis spectrophotometer to show the solution wavelengths.

The color change qualitatively indicates the photocatalytic activity of the sample coated by TiO$_2$ nanoparticles. The UV-Vis absorption spectroscopy device was used for the quantitative evaluation of the photocatalytic properties of the sample. The UV-Vis spectrophotometer showed the highest amount of absorption at the wavelengths of 288 and 351 nm which represent the presence of I$_2$ and I$_3^-$ atoms, respectively. To check the durability of TiO$_2$ nanoparticles, the KI test may be done under UV radiation in different periods. In the solution containing the unmodified membrane, no change of color and no absorption occurred after 6 h of UV radiation, while in the solution containing the membrane coated with TiO$_2$ nanoparticles both color change to yellow and absorption peak were observed. As is shown in Figure 7 absorption in each period for both samples remained relatively unchanged. An unchanged absorption spectrum during the period not only reflects its excellent photocatalytic activity, but is also proof of the suitable durability of TiO$_2$ coatings.

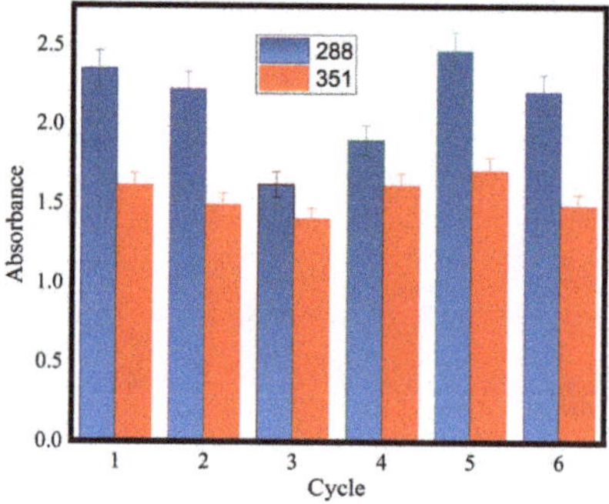

Figure 7. The absorption diagram of the KI solution containing the TiO$_2$–PVDF membrane at the wavelengths of 351 and 288 nm after it was exposed to UV for 6 h and for 6 periods.

3.3. Membrane Hydrophobicity

As was mentioned above, in this research, the hydrophobic properties of the membrane were optimized in two stages: first, changing the topography of the surface to enhance roughness by coating TiO_2 nanoparticles on the surface of the membrane; second, changing the chemical properties of the surface to repel water more using the functional agent FTCS. Then the modified membrane was used in a DCMD membrane module at the experimental scale.

Although the coating of TiO_2 nanoparticles on the surface of the PVDF membrane can reduce its hydrophobic properties due to the hydrophilic properties of TiO_2, in this study, the purpose of using titanium oxide nanoparticles was to create roughness and the hydrophilic modification of the membrane was not the aim. A multi-layer roughness can change the wettability properties of a membrane [34,35]. An increase in surface roughness (after the surface is coated with TiO_2 nanoparticles and a multi-layer roughness is formed on it) can lead to a strong capillary water suction on the membrane surface which expands the droplet on the surface until saturation [36]. Therefore, the hydrophilic properties of the surface increase.

Then the PVDF membrane was coated by FTCS and a substantial reduction was observed in the hysteresis angle from 18° for the unmodified membrane to about 6° for the FTCS–PVDF membrane. The increase in the water contact angle for the FTCS–PVDF membrane surface may be because of the notable reduction of the free energy of the membrane surface and its increased roughness.

Hydrophobic surfaces repel water and an air gap is created between the membrane and water droplets. This gap decreases the wettability of the membrane surface. In this case, water droplets can only have contact with the tip of the roughness; therefore, the physical contact between water droplets and the surface is considerably reduced. This is due to the fact that an air gap is created among the roughness bumps and surface tensile forces. Water droplets on superhydrophobic surfaces have small hysteresis contact angles and lie on the membrane surface spherically with a good approximation. Thus, they can roll even on surfaces with low slopes and remove pollutants from the surface (called the 'self-cleaning' property).

To increase the hydrophobic properties of the membrane, the TiO_2–PVDF membrane was coated by FTCS. The unmodified PVDF membrane had the contact angle of 89° ± 8° degrees with water. It is worth noting that the surface modification (FTCS–TiO_2–PVDF) created a contact angle of about 174° ± 10°. The unmodified and TiO_2–PVDF membranes lacked self-cleaning properties on steep slopes, while the FTCS–TiO_2–PVDF membrane showed self-cleaning properties even at low slopes.

3.4. Membrane Distillation at the Experimental Scale

Tests with a 5 wt % potassium nitrate solution were conducted for the unmodified and modified membranes. When the unmodified membrane was used, the cross-membrane water flux was 2×10^{-3} Kg/m^{2}s and when the FTCS–TiO_2–PVDF membrane was employed, the cross-membrane water flux was 6.4×10^{-4} Kg/m^{2}s. When each of the membranes was used, the electrical conductivity of the permeate was zero. In general, flux reduction was observed in both cases over time which could have two reasons. The first reason may be the blockage or fouling of the membrane pores and the reduced surface area available for the passage of water vapor; the second reason could be the decrease of the driving force (i.e., the reduction of temperature difference between the feed and permeate) [37]. To remove the fouling of the pores, each membrane was placed for 15 min in a NaOH 0.2 wt % solution with the pH of 12 at room temperature (25 °C). After being washed and reused, the modified membrane had the recovered flux of up to 95% and the unmodified membrane had the recovered flux of about 60%. After the membranes were washed, colors and sediments were removed from the modified membrane, while in the unmodified membrane, the color change was not resolved requiring more time for washing and sometimes backwashing. This phenomenon can show the anti-sedimentation property of the modified membrane. In the current research, the coat is easy to produce due to the low production temperature and the easy methods selected for coating. Hence, in case of damage to the coat of the membrane during washing, the coat can be easily restored.

3.5. Physical and Chemical Characteristics of the Permeate

As can be observed in Figure 8a, the electrical conductivity of the feed had an ascending trend and increased from 200 to 244 μs during a 6 h continuous test. Nevertheless, the electrical conductivity of the permeate remained constant throughout the test and was equal to that of distilled water. This result indicates that no ion was able to penetrate to the other side of the membrane and the permeate was pure.

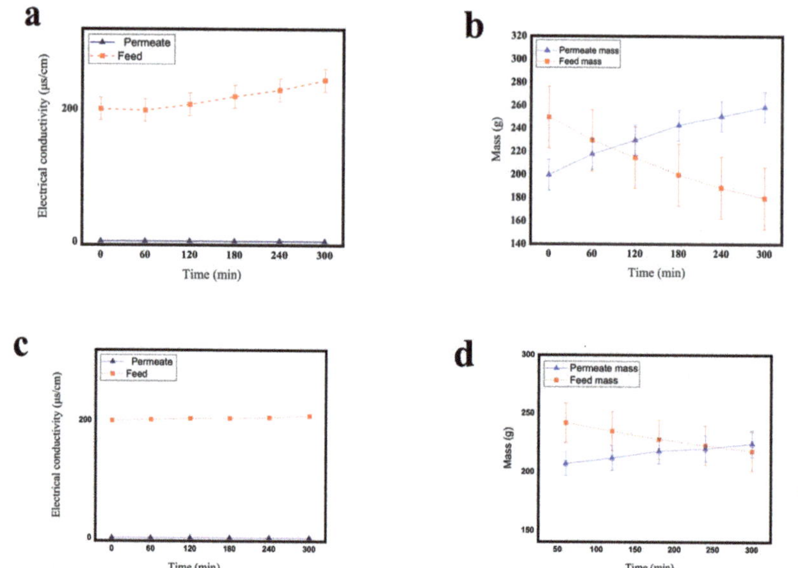

Figure 8. (**a**) The electrical conductivity-time diagram of the membrane distillation process using the PVDF membrane; (**b**) Mass-time diagram of the membrane distillation process using the PVDF membrane; (**c**) Electrical conductivity-time diagram of the membrane distillation process using the FTCS–TiO$_2$–PVDF membrane; (**d**) Mass-time diagram of the membrane distillation process using the FTCS–TiO$_2$–PVDF membrane.

Moreover, the mass of the feed had a descending trend and the mass of the permeate had an ascending trend indicating the transfer of mass through the membrane from the feed side to the permeate side Figure 8b.

In what follows, the above experiment was conducted using the FTCS–TiO$_2$–PVDF membrane. As can be seen in Figure 8c, during 6 h of continuous testing, the electrical conductivity of the feed had an ascending trend and increased from 200 to 208.6 μs, while the electrical conductivity of the permeate remained constant during the experiment and was equal to that of distilled water.

Moreover, the mass of the feed had a descending trend and the mass of the permeate had an ascending trend indicating the transfer of mass through the membrane from the feed side to the permeate side Figure 8d.

Comparing Figure 8a,c shows that the slope of the feed of the PVDF membrane is higher than that of the FTCS–TiO$_2$–PVDF membrane which may be due to the higher flux in the PVDF membrane. Interestingly, the slope of the permeate is almost constant in Figure 8b,d, indicating the purity of the permeate.

The interesting point about the performance of the unmodified and modified membranes in the two tests mentioned above is the difference between the increased electrical conductivity of the feed in them over 6 h. According to the mass-time diagrams, this difference could be due to the lower mass reduction of the modified membrane Figure 8b than that of the unmodified membrane Figure 8d. There are detailed explanations regarding mass transfer reduction. As can be observed in

Figure 8d, the flux is not as expected which is a disadvantage of this system and needs to be improved. Future studies may focus on solving this problem.

3.6. Process Simulation Using MATLAB at Pilot Scale

At this point, using the MATLAB software [33], the obtained results for the modified membrane at the laboratory scale were simulated at the pilot scale as shown in Table 1.

Table 1. Comparing the results of the operational parameters at the pilot and laboratory scales.

Results of Operational Parameters	Lab Scale	Pilot Scale
Feed-permeate temperature ($°C$)	77–12.30	48.33–21.62
KNO_3 concentration (g/kg)	0.9	35
Feed-permeate outlet mass flow rate (kg/s)	$(3.2–0.56) \times 10^{-3}$	0.99–1
Cross-membrane flux ($Kgm^{-2}h^{-1}$)	2.3	0.96
Effective membrane area (m^2)	14×10^{-4}	0.5

What is noticeable in the simulation is the low temperature of the feed (below 50 $°C$) at the pilot scale. Another point to note is the ability of this method to remove high concentrations of nitrate at the pilot scale. According to the data obtained from the simulation, a small area of the modified membrane (0.5 m^2) is enough to remove nitrate from water with the concentration of 35 g/kg in a low temperature (48.33 $°C$) with the flow rate of 1 Kg/s and cross-membrane flux (0.96 $Kgm^{-2}h^{-1}$). The reason for the reduction of flux at the pilot scale may be the decrease of the driving force temperature difference). With the increase of temperature difference, the flux will probably increase (Figure S3; Video S1).

4. Conclusions

The PVDF membrane was coated in two steps with TiO_2 nanoparticles and 1H,1H, 2H,2H-Perfluorododecyltrichlorosilane. The hierarchical structure of the membrane with a multilayer roughness resulted in an increase in the contact angle of the membrane from 89° ± 8° to 174° ± 10°. The membrane performance was evaluated by the direct contact membrane distillation method using 5 wt % potassium nitrate at the laboratory scale. The electrical conductivity of the feed containing nitrate increased, while the electrical conductivity of the permeate remained constant during the entire process. Furthermore, the low temperature and high flux at the simulated pilot-scale indicate that this method is highly efficient.

At the pilot-scale, using the membrane distillation method and 0.5 m^2 of the modified membrane, it is possible to treat water polluted with nitrate with the concentration of 35 g/Kg at the low temperature of 48.33 $°C$ and the flow rate of 1 Kg/s with the cross-membrane flux (0.96 $Kgm^{-2}h^{-1}$).

Supplementary Materials: The following are available online at http://www.mdpi.com/2073-4360/12/12/2774/s1, Figure S1: Lab scale setup of Direct Contact Membrane Distillation, Table S1: List of input conditions to software, Table S2: List of the output conditions from running the application, Figure S2: 4. Simulation output at pilot scale with PVDF-TiO2 -FTCS membrane; Figure S3: Schematic presentation of membrane distillation by superhydrophobic membrane. Video S1: Live cover.

Author Contributions: Conceptualization, A.R. and Y.O.; Experiments and analyses, F.E.; writing—original draft preparation F.E.; writing—review and editing, Y.O. and A.R.; funding acquisition, Y.O. and A.R. All authors have read and agreed to the published version of the manuscript.

Funding: This research was funded by grant numbers 163020139, 164020247 and 163020211.

Acknowledgments: The authors hereby express their thanks for the support rendered by University of Isfahan and Nanjing Forestry University.

Conflicts of Interest: The authors declare no conflict of interest.

References

1. Karimi-Maleh, H.; Arotiba, O.A. Simultaneous determination of cholesterol, ascorbic acid and uric acid as three essential biological compounds at a carbon paste electrode modified with copper oxide decorated reduced graphene oxide nanocomposite and ionic liquid. *J. Colloid Interface Sci.* **2020**, *560*, 208–212. [CrossRef]

2. Karimi-Maleh, H.; Sheikhshoaie, M.; Sheikhshoaie, I.; Ranjbar, M.; Alizadeh, J.; Maxakato, N.W.; Abbaspourrad, A. A novel electrochemical epinine sensor using amplified CuO nanoparticles and a n-hexyl-3-methylimidazolium hexafluorophosphate electrode. *New J. Chem.* **2019**, *43*, 2362–2367. [CrossRef]

3. Karimi-Maleh, H.; Shojaei, A.F.; Tabatabaeian, K.; Karimi, F.; Shakeri, S.; Moradi, R. Simultaneous determination of 6-mercaptopruine, 6-thioguanine and dasatinib as three important anticancer drugs using nanostructure voltammetric sensor employing Pt/MWCNTs and 1-butyl-3-methylimidazolium hexafluoro phosphate. *Biosens. Bioelectron.* **2016**, *86*, 879–884. [CrossRef] [PubMed]

4. Khodadadi, A.; Faghih-Mirzaei, E.; Karimi-Maleh, H.; Abbaspourrad, A.; Agarwal, S.; Gupta, V.K. A new epirubicin biosensor based on amplifying DNA interactions with polypyrrole and nitrogen-doped reduced graphene: Experimental and docking theoretical investigations. *Sens. Actuators B Chem.* **2019**, *284*, 568–574. [CrossRef]

5. Miraki, M.; Karimi-Maleh, H.; Taher, M.A.; Cheraghi, S.; Karimi, F.; Agarwal, S.; Gupta, V.K. Voltammetric amplified platform based on ionic liquid/NiO nanocomposite for determination of benserazide and levodopa. *J. Mol. Liq.* **2019**, *278*, 672–676. [CrossRef]

6. Tahernejad-Javazmi, F.; Shabani-Nooshabadi, M.; Karimi-Maleh, H. Analysis of glutathione in the presence of acetaminophen and tyrosine via an amplified electrode with MgO/SWCNTs as a sensor in the hemolyzed erythrocyte. *Talanta* **2018**, *176*, 208–213. [CrossRef]

7. Karimi-Maleh, H.; Cellat, K.; Arıkan, K.; Savk, A.; Karimi, F.; Şen, F. Palladium–Nickel nanoparticles decorated on Functionalized-MWCNT for high precision non-enzymatic glucose sensing. *Mater. Chem. Phys.* **2020**, *250*. [CrossRef]

8. Orooji, Y.; Ghasali, E.; Emami, N.; Noorisafa, F.; Razmjou, A. ANOVA design for the optimization of TiO_2 coating on polyether sulfone membranes. *Molecules* **2019**, *24*, 2924. [CrossRef] [PubMed]

9. Orooji, Y.; Liang, F.; Razmjou, A.; Li, S.; Mofid, M.R.; Liu, Q.; Guan, K.; Liu, Z.; Jin, W. Excellent biofouling alleviation of thermoexfoliated vermiculite blended poly (ether sulfone) ultrafiltration membrane. *ACS Appl. Mater. Interfaces* **2017**, *9*, 30024–30034. [CrossRef] [PubMed]

10. Orooji, Y.; Liang, F.; Razmjou, A.; Liu, G.; Jin, W. Preparation of anti-adhesion and bacterial destructive polymeric ultrafiltration membranes using modified mesoporous carbon. *Sep. Purif. Technol.* **2018**, *205*, 273–283. [CrossRef]

11. Razmjou, A.; Eshaghi, G.; Orooji, Y.; Hosseini, E.; Korayem, A.H.; Mohagheghian, F.; Boroumand, Y.; Noorbakhsh, A.; Asadnia, M.; Chen, V. Lithium ion-selective membrane with 2D subnanometer channels. *Water Res.* **2019**, *159*, 313–323. [CrossRef] [PubMed]

12. Razmjou, A.; Asadnia, M.; Hosseini, E.; Korayem, A.H.; Chen, V. Design principles of ion selective nanostructured membranes for the extraction of lithium ions. *Nat. Commun.* **2019**, *10*, 5793. [CrossRef] [PubMed]

13. Gu, B.; Ge, Y.; Chang, S.X.; Luo, W.; Chang, J. Nitrate in groundwater of China: Sources and driving forces. *Glob. Environ. Chang.* **2013**, *23*, 1112–1121. [CrossRef]

14. Zhai, Y.; Lei, Y.; Zhou, J.; Li, M.; Wang, J.; Teng, Y. The spatial and seasonal variability of the groundwater chemistry and quality in the exploited aquifer in the Daxing District, Beijing, China. *Environ. Monit. Assess.* **2015**, *187*. [CrossRef] [PubMed]

15. Razmjou, A.; Hosseini, E.; Cha-Umpong, W.; Korayem, A.H.; Asadnia, M.; Moazzam, P.; Orooji, Y.; Karimi-Maleh, H.; Chen, V. Effect of chemistry and geometry of GO nanochannels on the Li ion selectivity and recovery. *Desalination* **2020**, *496*. [CrossRef]

16. Goudarzi, S.; Jozi, S.A.; Monavari, S.M.; Karbasi, A.; Hasani, A.H. Assessment of groundwater vulnerability to nitrate pollution caused by agricultural practices. *Water Qual. Res. J.* **2017**, *52*, 64–77. [CrossRef]

17. Kazakis, N.; Voudouris, K.S. Groundwater vulnerability and pollution risk assessment of porous aquifers to nitrate: Modifying the DRASTIC method using quantitative parameters. *J. Hydrol.* **2015**, *525*, 13–25. [CrossRef]

18. Shrestha, S.; Semkuyu, D.J.; Pandey, V.P. Assessment of groundwater vulnerability and risk to pollution in Kathmandu Valley, Nepal. *Sci. Total Environ.* **2016**, *556*, 23–35. [CrossRef]

19. Rahmati, O.; Melesse, A.M. Application of Dempster–Shafer theory, spatial analysis and remote sensing for groundwater potentiality and nitrate pollution analysis in the semi-arid region of Khuzestan, Iran. *Sci. Total Environ.* **2016**, *568*, 1110–1123. [CrossRef]

20. Gao, Y.; Yu, G.; Luo, C.; Zhou, P. Groundwater nitrogen pollution and assessment of its health risks: A case study of a typical village in rural-urban continuum, China. *PLoS ONE* **2012**, *7*. [CrossRef]

21. Johnson, S.F. Methemoglobinemia: Infants at risk. *Curr. Probl. Pediatr. Adolesc. Health Care* **2019**, *49*, 57–67. [CrossRef] [PubMed]

22. Wongsanit, J.; Teartisup, P.; Kerdsueb, P.; Tharnpoophasiam, P.; Worakhunpiset, S. Contamination of nitrate in groundwater and its potential human health: A case study of lower Mae Klong river basin, Thailand. *Environ. Sci. Pollut. Res.* **2015**, *22*, 11504–11512. [CrossRef] [PubMed]

23. Rahmati, O.; Samani, A.N.; Mahmoodi, N.; Mahdavi, M. Assessment of the contribution of N-fertilizers to nitrate pollution of groundwater in western Iran (Case Study: Ghorveh–Dehgelan Aquifer). *Water Qual. Expo. Health* **2015**, *7*, 143–151. [CrossRef]

24. Pisciotta, A.; Cusimano, G.; Favara, R. Groundwater nitrate risk assessment using intrinsic vulnerability methods: A comparative study of environmental impact by intensive farming in the Mediterranean region of Sicily, Italy. *J. Geochem. Explor.* **2015**, *156*, 89–100. [CrossRef]

25. Tatarczak-Michalewska, M.; Flieger, J.; Kawka, J.; Płaziński, W.; Flieger, W.; Blicharska, E.; Majerek, D. HPLC-DAD Determination of Nitrite and Nitrate in Human Saliva Utilizing a Phosphatidylcholine Column. *Molecules* **2019**, *24*, 1754. [CrossRef] [PubMed]

26. Joseph, A.C. 2017 WHO Guidelines for drinking-water quality: First addendum to the fourth edition. *J. AWWA* **2017**. [CrossRef]

27. Boubakri, A.; Hafiane, A.; Bouguecha, S.A. Nitrate removal from aqueous solution by direct contact membrane distillation using two different commercial membranes. *Desalin. Water Treat.* **2015**, *56*, 2723–2730. [CrossRef]

28. Arefi-Oskoui, S.; Khataee, A.; Safarpour, M.; Orooji, Y.; Vatanpour, V. A review on the applications of ultrasonic technology in membrane bioreactors. *Ultrason. Sonochem.* **2019**, *58*. [CrossRef]

29. Razmjou, A.; Arifin, E.; Dong, G.; Mansouri, J.; Chen, V. Superhydrophobic modification of TiO_2 nanocomposite PVDF membranes for applications in membrane distillation. *J. Membr. Sci.* **2012**, *415*, 850–863. [CrossRef]

30. Zhang, J.; Song, Z.; Li, B.; Wang, Q.; Wang, S. Fabrication and characterization of superhydrophobic poly (vinylidene fluoride) membrane for direct contact membrane distillation. *Desalination* **2013**, *324*, 1–9. [CrossRef]

31. Chaharmahali, A.R. The Effect of TiO_2 Nanoparticles on the Surface Chemistry, Structure and Fouling Performance of Polymeric Membranes. Ph.D. Thesis, University of New South Wales, Sydney, Australia, 2012.

32. Dong, G.; Kim, J.F.; Kim, J.H.; Drioli, E.; Lee, Y.M. Open-source predictive simulators for scale-up of direct contact membrane distillation modules for seawater desalination. *Desalination* **2017**, *402*, 72–87. [CrossRef]

33. Hamzah, N.; Leo, C.; Ooi, B. Superhydrophobic PVDF/TiO 2-SiO 2 Membrane with Hierarchical Roughness in Membrane Distillation for Water Recovery from Phenolic Rich Solution Containing Surfactant, Chinese. *J. Polym. Sci.* **2019**, *37*, 609–616.

34. Qing, W.; Wang, J.; Ma, X.; Yao, Z.; Feng, Y.; Shi, X.; Liu, F.; Wang, P.; Tang, C.Y. One-Step Tailoring Surface Roughness and Surface Chemistry to Prepare Superhydrophobic Polyvinylidene Fluoride (PVDF) Membranes for Enhanced Membrane Distillation Performances. *J. Colloid Interface Sci.* **2019**, *553*, 99–107. [CrossRef] [PubMed]

35. Gupta, P.; Kandasubramanian, B. Directional fluid gating by janus membranes with heterogeneous wetting properties for selective oil–water separation. *ACS Appl. Mater. Interfaces* **2017**, *9*, 19102–19113. [CrossRef]

36. Goh, S.; Zhang, Q.; Zhang, J.; McDougald, D.; Krantz, W.B.; Liu, Y.; Fane, A.G. Impact of a biofouling layer on the vapor pressure driving force and performance of a membrane distillation process. *J. Membr. Sci.* **2013**, *438*, 140–152. [CrossRef]

37. Chew, J.W.; Krantz, W.B.; Fane, A.G. Effect of a macromolecular-or bio-fouling layer on membrane distillation. *J. Membr. Sci.* **2014**, *456*, 66–76. [CrossRef]

Publisher's Note: MDPI stays neutral with regard to jurisdictional claims in published maps and institutional affiliations.

Article

3D Nanoarchitecture of Polyaniline-MoS$_2$ Hybrid Material for Hg(II) Adsorption Properties

Hilal Ahmad [1,2], Ibtisam I. BinSharfan [3], Rais Ahmad Khan [3] and Ali Alsalme [3,*]

[1] Division of Computational Physics, Institute for Computational Science, Ton Duc Thang University, Ho Chi Minh City 700000, Vietnam; hilalahmad@tdtu.edu.vn
[2] Faculty of Applied Sciences, Ton Duc Thang University, Ho Chi Minh City 700000, Vietnam
[3] Department of Chemistry, College of Science, King Saud University, Riyadh 11451, Saudi Arabia; ibtisam.i.sh@hotmail.com (I.I.B.); krais@ksu.edu.sa (R.A.K.)
* Correspondence: aalsalme@ksu.edu.sa

Received: 25 October 2020; Accepted: 16 November 2020; Published: 17 November 2020

Abstract: We report the facile hydrothermal synthesis of polyaniline (PANI)-modified molybdenum disulfide (MoS$_2$) nanosheets to fabricate a novel organic–inorganic hybrid material. The prepared 3D nanomaterial was characterized by field emission scanning electron microscopy, high-resolution transmission electron microscopy, energy-dispersive X-ray spectroscopy and X-ray diffraction studies. The results indicate the successful synthesis of PANI–MoS$_2$ hybrid material. The PANI–MoS$_2$ was used to study the extraction and preconcentration of trace mercury ions. The experimental conditions were optimized systematically, and the data shows a good Hg(II) adsorption capacity of 240.0 mg g^{-1} of material. The adsorption of Hg(II) on PANI–MoS$_2$ hybrid material may be attributed to the selective complexation between the–S ion of PANI–MoS$_2$ with Hg(II). The proposed method shows a high preconcentration limit of 0.31 μg L^{-1} with a preconcentration factor of 640. The lowest trace Hg(II) concentration, which was quantitatively analyzed by the proposed method, was 0.03 μg L^{-1}. The standard reference material was analyzed to determine the concentration of Hg(II) to validate the proposed methodology. Good agreement between the certified and observed values indicates the applicability of the developed method for Hg(II) analysis in real samples. The study suggests that the PANI–MoS$_2$ hybrid material can be used for trace Hg(II) analyses for environmental water monitoring.

Keywords: toxicity; polyaniline; mercury; adsorption; MoS$_2$

---

1. Introduction

Mercury (Hg(II)) is one of the most toxic metal pollutants found in the environment and ranks third after arsenic and lead in the National Priorities List of the Agency for Toxic Substances and Disease Registry (ATSDR) [1–3]. The Hg(II) contamination of ground and surface water results from geochemical reactions and anthropogenic activities such as improper dumping of electronic waste, thermometer, barometer and mercury lamp waste. Human exposure to metal ions, including Hg(II), can occur during occupational activities, mainly through inhalation and dermal routes in mining and industry, and over a lifetime, from water and food consumption and exposure to soil, dust and air [4,5]. Long-term consumption of drinking water contaminated with Hg(II) can be associated with increased risk of cancers, reproductive problems, detrimental effects on the human brain, blood circulation, immune and reproductive systems and cardiovascular disease [2,6,7]. Therefore, to minimize these risks, the United States Environmental Protection Agency (USEPA) has set the maximum permissible limit of 2 μg L^{-1} [8].

Modern analytical techniques such as X-ray fluorescence, atomic absorption spectrometry, inductively coupled plasma atomic emission spectrometry, and inductively coupled plasma mass spectrometry have been widely used for the analysis of Hg(II) [9–11]; however, direct determination of Hg(II) in real aqueous samples is challenging due to their low concentrations and complexity of sample matrices [12]. Therefore, preliminary extraction and preconcentration steps are often necessary before instrumental determination. Various separation methods such as solvent extraction, hydride generation, electro-coagulation, precipitation, cloud point extraction and solid-phase extraction (SPE) are employed to extract metal ions [13–17]. SPE is a preferred procedure because of its advantages such as easy operation, the negligible use of organic solvents, complete desorption of analytes, high preconcentration factor, and used in both batch and column modes [18,19]. Adsorption of the analyte onto nanomaterials in SPE is considered an efficient process based on factors like the high surface area of sorbent, efficient adsorption capacity, and easy functionalize activity [20–23]. Nanomaterial-based adsorbents have been extensively researched in the past two decades to find new solutions or to enhance the existing solutions in environmental water remediation [21,24–26]. In recent years, two-dimensional (2D) nanostructures such as metal chalcogenides, metal hydroxides, and double-layered metal hydroxides have attracted tremendous interest due to their high surface area and a porous structure with large surface active sites [27–32]. However, the critical drawback of directly employing these 2D materials in the SPE column is its small size and dispersion in aqueous media, leading to loss of adsorbent during a column operation. Moreover, for the effective deployment of 2D nanostructures, they must prevent stacking. The weak interlayer bonding and low free spacing cause the stacking of nanosheets in the SPE column.

In the present work, we fabricate a blend of 3D hybrid material (organic–inorganic composite) made from 2D MoS_2 and a 1D polymer polyaniline (PANI) via in situ oxidative polymerization of PANI with exfoliated MoS_2 nanosheets to overcome the limitations mentioned above. The integration of MoS_2 nanosheets with PANI restricts the nanosheets leaching from the column and provide stability in aqueous media. Wang et al. reported the polyaniline/zirconium composite to remove organic pollutants [33]. Similarly, Gao et al. reported the hybrid polyaniline/titanium phosphate composite to remove Re(VII) [34]. Moreover, there are no reports on Hg (II) extraction using PANI–MoS_2 hybrid material. The extensive and profound studies are carried out using PANI–MoS_2 hybrid nanomaterial to develop a column SPE method for the extraction of trace Hg(II). The accuracy and applicability of the developed method were validated by analyzing the certified reference material and by spiking of real environmental water samples.

2. Experimental Details

2.1. Materials and Methods

2.1.1. Chemicals and Reagents

All the chemicals used were of analytical grade. Sodium molybdate ($Na_2MoO_4 \cdot 2H_2O$), ammonium persulfate ((NH_4)2S_2O_8) and hydrochloric acid were purchased from Thermo Fisher Scientific, New Delhi, India. Thioacetamide (C_2H_5NS), silicotungstic acid AR [$H(Si(W_3O_{10})_4) \cdot xH_2O$] and aniline ($C_6H_5NH_2$), with 99% purity, were purchased from Sigma Aldrich (Steinem, Germany). A stock solution of divalent mercury ions ($Hg(NO_3)_2$) of 1000 mg L^{-1} was bought from Agilent (Melbourne, Australia) and used after successive dilutions. A 1 M of HNO_3 and NaOH solution was used to adjust the sample pH.

2.1.2. Synthesis of PANI–MoS_2 Hybrid Material

The PANI–MoS_2 hybrid material was synthesized in two steps. In the first step, MoS_2 nanosheets were hydrothermally synthesized. Briefly, 0.2 mM of sodium molybdate, 1.8 mM of thioacetamide and 5.6 mM of silicotungstic acid were dissolved in 100 mL of deionized water. The reaction mixture was kept in 250 mL of Teflon-coated hydrothermal assembly and heated at 220 °C for 24 h using an air oven. The obtained MoS_2 nanosheets (0.2 g) were ultrasonicated using probe sonicator in 20 mL of deionized water for 40 min at 27 °C. In the second step, the in situ oxidative polymerization of aniline monomers was carried out onto presynthesized MoS_2 nanosheets using ammonium persulfate oxidizer. In this process, 4 mL of aniline monomer, 6 mL of HCl and 40 mL of deionized water were stirred together and refrigerated for three h. The cooled reaction solution was added to the exfoliated (ultrasonicated) MoS_2 nanosheets solution. The formed suspension was stirred in an ice bath ($-5\ °C$) for 30 min. Finally, 10 mL of ammonium persulfate (0.2 M) was added dropwise in the suspension and continuously stirred for 3 h. The obtained solution was filtered, and the residue was washed with deionized water and ethanol. The residue was dried in a vacuum at 60 °C for 12 h. The obtained PANI–MoS_2 hybrid material was characterized and studied for Hg(II) adsorption properties.

2.2. Material Characterization

The surface morphology and structural properties were observed using a scanning electron microscope (FE-SEM, Zeiss, Sigma, Tokyo, Japan) and high-resolution transmission electron microscopy (HR-TEM F30 S-Twin TECNAI FEI, Tokyo, Japan) operating at an acceleration voltage of 300 kV. Samples for HRTEM characterization were prepared by dispersing the material powder into ethanol by ultrasonic treatment. Rigaku Smart Lab X-ray diffractometer with Cu K_α radiation at 1.540 Å in the 2θ range of 20–90° is used to study crystal structure and phase determination. The Brunauer–Emmett–Teller (BET) surface area measurements were carried out using an Autosorb-iQ one-station (Quantachrome Instruments, Boynton Beach, FL, USA). The nitrogen gas was used for sorption and desorption analysis at low relative pressures. The surface charge of the materials was investigated by Zeta potential (z) measurements on a Zetasizer (Malvern Instruments, Malvern, UK). A Shimadzu TGA-50 thermal analyzer was used to conduct thermal gravimetric analysis (TGA) at a heating rate of 10 °C/min from 27 °C to 650 °C. A Perkin Elmer inductively coupled plasma optical emission spectrometer (ICP-OES model Avio 200, Melbourne, Australia) was used to analyze the Hg(II) concentrations. The ATR-IR (attenuated total reflectance infrared spectroscopy) (Vertex 70v, Bruker, Ettlingen, Germany) analysis of PANI–MoS_2 adsorbent, before and after Hg(II) adsorption, were carried out in the range of 400–4000 cm^{-1} (with the accumulation of 60 scans). The surface elemental analysis was carried out using X-ray photoelectron spectroscopy (XPS, Thermo Fisher Scientific ESCALAB 250Xi, Waltham, MA, USA). The studies were performed in a binding energy range of 0–1400 eV. MgK alpha was used as an X-ray source at 1253.6 eV with a detection angle of 45° and a depth of 10 nm.

2.3. Recommended Column Procedure

A polytetrafluoroethylene column (Length = 10 cm; diameter = 1 cm) (Merck, Shanghai, China) packed with 0.5 g of PANI–MoS_2 hybrid material (bed height = 1.6 cm) was used for the column through experiments. A bench of model solutions (100 mL) of desired Hg(II) concentration maintained at pH 6.0 using 1 M of HNO_3 and NaOH solution were percolated through the column bed at a flow rate of 8 mL min^{-1} using a peristaltic pump (Scenchen, Hebei, China). The adsorbed Hg(II) was stripped out using a 5 mL of 0.5 M HCl, and the concentration of adsorbed Hg(II) was analyzed by ICP-OES.

3. Results and Discussion

3.1. Characterization

The surface morphology of MoS_2 and PANI–MoS_2 hybrid composite is shown in Figure 1A,B. Figure 1A shows the MoS_2 nanosheets arranged in a flower-like structure with porous morphology. Figure 1B shows that PANI uniformly bounded the MoS_2 sheets. The resulting PANI–MoS_2 structure had a long tube-like morphology with a rough surface due to constituted nanoparticles, indicating that the PANI–MoS_2 may provide additional binding sites for Hg(II) adsorption. The difference in the HRTEM images of Figure 1C,D reveals that the PANI was successfully immobilized on MoS_2 nanosheets. From Figure 1C,D, the two contrasted regions, the dark region representing MoS_2 nanosheets, nearby many ultrathin single MoS_2 nanosheets, were also present, and the lighter region represents PANI nanofibers. Figure 2A,B illustrates the SEM and EDX spectra of PANI–MoS_2 after Hg(II) adsorption. Figure 3A,B shows the X-ray diffraction (XRD) pattern of MoS_2 and $PANI–MoS_2$. The diffraction peaks observed at 2θ = 13.10, 32.70, 35.15, 41.50 and 59.50 corresponds to the (002), (100), (103), (015) and (110) planes of MoS_2 (Figure 3A). The d-spacing of MoS_2 calculated using Bragg's law was found to be 6.71 Å. From the XRD data (Figure 3B), the interlayer spacing of MoS_2 nanosheets in $PANI–MoS_2$ hybrid material was found to be 6.24 Å. The observed data depicted that the aniline forms mono and bilayers structures on MoS_2 and the polymerization of intercalated aniline monomer reduces the interlayer distance from 6.71 to 6.24 Å. It was suggested that the polymerization of aniline occurs outside the MoS_2 nanosheets. Also, the $PANI–MoS_2$ hybrid material was less crystalline than bare MoS_2 attributes to flexible $PANI–MoS_2$ hybrid structure with an amorphous surface. The nitrogen gas adsorption–desorption analysis was carried out to characterize the physical properties of the adsorbent; the nitrogen isotherms are shown in Figure 4. The average surface area calculated by the Brunauer–Emmett–Teller (BET) method was found to be 29.0 $m^2\,g^{-1}$. The thermal analysis of $PANI–MoS_2$ under air atmosphere was carried out to study the thermal stability. It was observed that the material has thermal stability, up to a temperature of 320 °C (Figure 5). The TGA shows minor weight loss around 100–120 °C, which may occur due to interlayer water content loss. The major weight loss commences at 320–600 °C may be attributed to the oxidative degradation of the polyaniline component of the $PANI–MoS_2$ hybrid material. The ATR-IR spectra of $PANI–MoS_2$ before and after Hg(II) adsorption is shown in Supplementary Materials Figure S1. The peaks observed at 1600, 1485, 1290 and 1150 cm^{-1} in the spectra of $PANI–MoS_2$ were attributed to the stretching vibrations of the C–C ring, C–H bending and C–N stretching vibrations of the quinoid and benzenoid ring of PANI, respectively. The characteristic MoS_2 peak was observed at 468 cm^{-1}. The small peak observed at 798 cm^{-1} was corresponds to S-S stretching vibration. After Hg(II) adsorption, the weak intensity peak observed at 450 cm^{-1} may be attributed to Hg-S stretching vibration. The elemental composition of $PANI–MoS_2$ was further examined by XPS analysis. Figure 6A,B shows the XPS survey of $PANI–MoS_2$ before and after Hg(II) adsorption. In Figure 6A, the peaks at binding energies of 162.0, 229.1, 285.0, 395.0 and 532 eV correspond to S 2p, Mo 3d, C 1s, N 1s and O 1s, respectively. In Figure 6B, the presence of Hg 4f peak at a binding energy of 100.6 eV attributes to the adsorption of Hg(II) onto $PANI–MoS_2$ adsorbent.

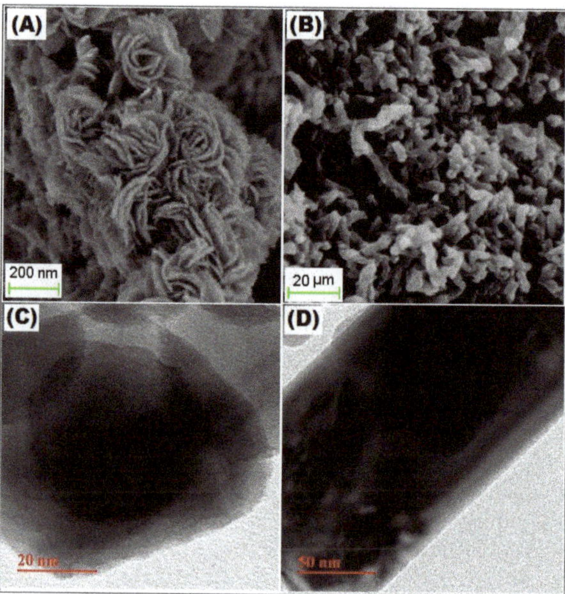

Figure 1. Scanning electron microscope image of (**A**) bare MoS_2; (**B**) polyaniline/molybdenum disulfide hybrid nanomaterial ($PANI-MOS_2$); and transmission electron microscopy image of (**C**) MoS_2; and (**D**) $PANI-MOS_2$.

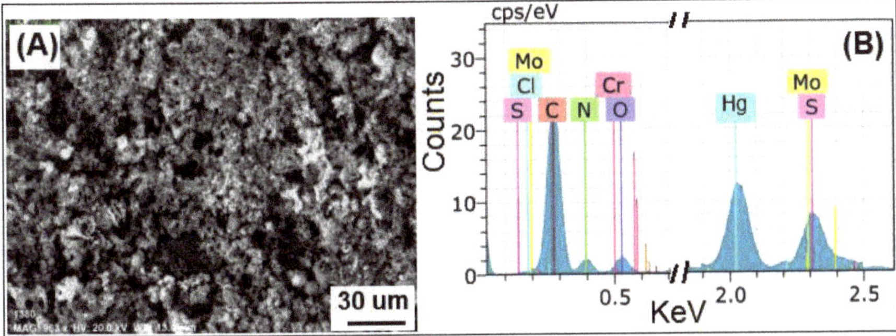

Figure 2. PANI–MoS$_2$ after Hg(II) adsorption (**A**) FESEM image (**B**) EDX spectra.

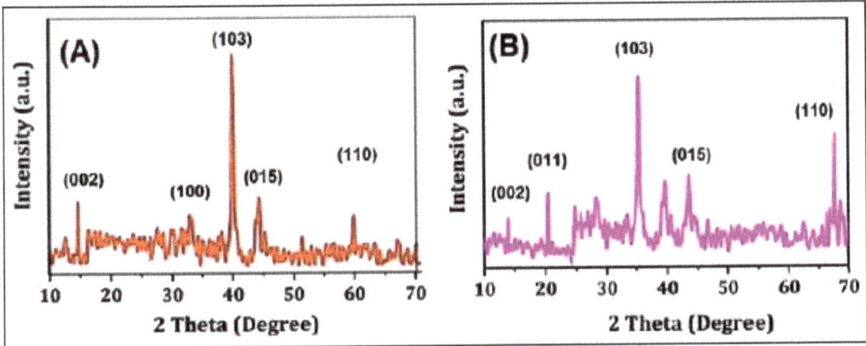

Figure 3. XRD diffraction pattern of (**A**) MoS$_2$; and (**B**) PANI-MoS$_2$.

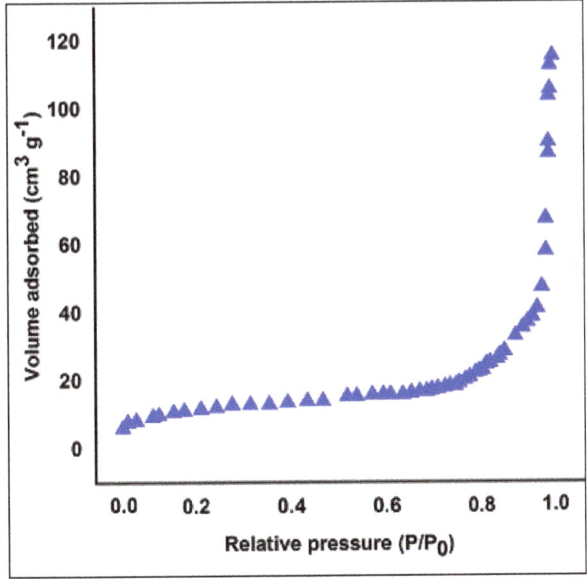

Figure 4. Nitrogen adsorption isotherm of PANI–MoS$_2$ adsorbent.

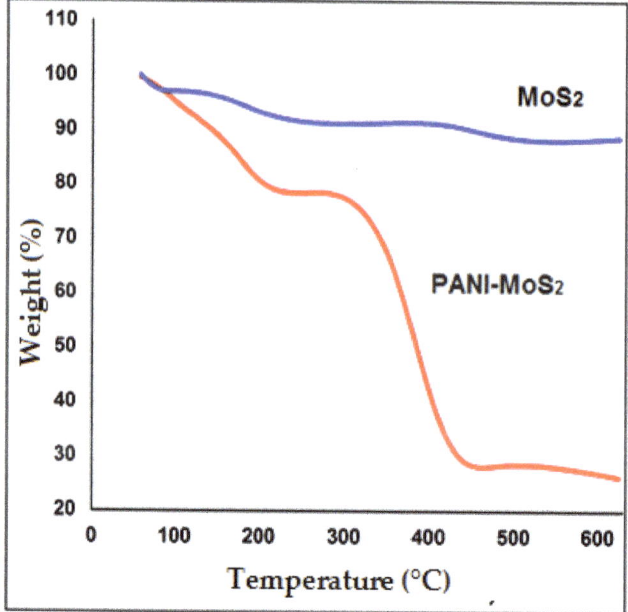

Figure 5. TGA profile of bare MoS_2 and PANI–MoS_2.

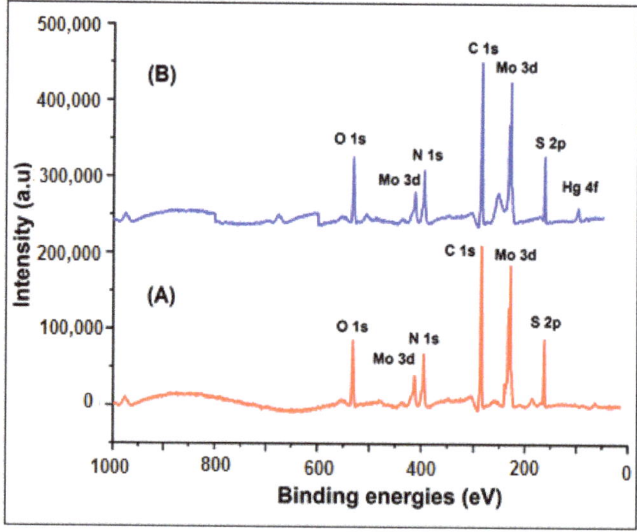

Figure 6. X-ray photoelectron spectra of PANI–MoS_2 (**A**) before Hg(II) adsorption and (**B**) after Hg(II) adsorption.

3.2. Optimized Sample pH and Adsorption Mechanism

The solution pH plays an essential role in the adsorption of the analyte by influencing the surface charge of adsorbent and metal ion species distribution. Optimum pH can reduce the interferences caused by the sample matrix and improves the method selectivity. Therefore, the optimization of sample pH is the first step. The adsorption of Hg(II) on PANI–MoS_2 was studied in the pH range of 1.0–7.0. Basic sample pH (pH 8.0 to 10.0) was avoided due to the formation of Hg precipitates.

A bench of model solutions (volume 100 mL), each containing 100 ppm of Hg(II) maintained at pH 1.0–7.0 (using 1 M of HNO_3 and NaOH solution), was passed through columns packed with 0.5 g of PANI–MoS_2 hybrid material. The adsorbed Hg(II) was eluted and subsequently determined by ICP-OES. As shown in Figure 7A, the PANI–MoS_2 hybrid material shows Hg(II) adsorption at a wider pH range. It can be seen that the Hg(II) adsorption at low pH values (up to pH 3) was not much affected and increased quickly after pH 4 and reached a maximum at pH 6.0–7.0. A complete recovery ca. 100% was observed at pH 6.0–7.0. The adsorption of Hg(II) mainly occurs on the active sites of PANI–MoS_2 composite via favorable chelation of Hg(II) with sulfide ions of PANI–MoS_2, in addition to the amine and imine functionalities of PANI. The intrinsic sulfur ions of PANI–MoS_2 hybrid material are the primary binding sites for the adsorption of Hg(II). At low pH values, the PANI–MoS_2 hybrid material shows less adsorption of Hg(II) due to the protonation of active/binding sites. At higher sample pH, the –S ions get deprotonated, and the soft-soft interaction between the -S ions and Hg(II) dominates thereby, increases the Hg(II) adsorption [35,36]. To better understand such observations, the surface charge of PANI–MoS_2 was measured (Figure 7B). For comparison, the zeta potential of nascent MoS_2 and PANI were also presented in Figure S2. The results of zeta potential indicate that at pH values 1.0–5.0, the PANI–MoS_2 surface was positively charged, resulting in weaker interaction between the surface groups and Hg(II) and above pH 5.0, the presence of negative charge on the surface of PANI–MoS_2 hybrid material, leading to the efficient adsorption of Hg(II) which is appropriate following the adsorption results (Figure 7A). In conclusion, the chelation of Hg(II) with the -S ions of PANI–MoS_2 hybrid material and the electrostatic interactions are the primary adsorption mechanisms for Hg(II); thus, pH 6.0 was chosen for the adsorption of Hg(II) in further experiments.

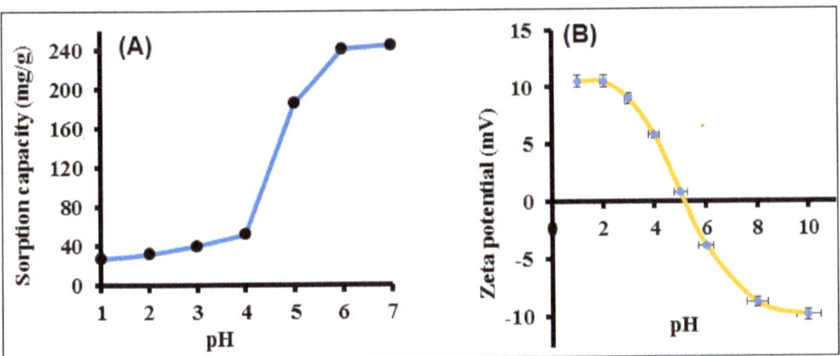

Figure 7. (**A**) Effect of sample pH on the adsorption of Hg(II); (**B**) zeta potential of PANI-MOS_2 adsorbent (experimental conditions: sorbent amount 0.5 g; sample volume 100 mL; flow rate 8 mL min^{-1}, Hg^{2+} 100 mg L^{-1}).

3.3. Preconcentration and Breakthrough Studies

Due to the ultra-low concentration of Hg(II) ions, direct instrumental determination of Hg(II) contamination level in surface and ground waters is challenging. Therefore, a preconcentration technique is a prerequisite to improve the analyte concentration by transforming it from a large sample volume to a smaller one. To analyze the preconcentration limit and preconcentration factor of the developed method, a series of model solutions with varying sample volume (1500–4000 mL), each contains a fixed amount of 1.0 µg of Hg(II) and maintained at pH 6.0, were passed through the column at a flow rate of 8 mL min^{-1}. The sorbed Hg(II) was then eluted using a suitable eluting agent, and the amount of Hg(II) was determined by ICP-OES. Table 1 illustrated the obtained results. It was observed that the quantitative recovery of Hg(II) was achieved within a sample volume of 3200 mL while on increasing the sample volume to 3500–4000 mL, the percent recovery of Hg(II) noticeably decreased to 90–85%. Thereby, a high preconcentration limit of 0.31 µg L^{-1} was obtained with a preconcentration factor of 640. Such a high preconcentration factor is necessitated for column preconcentration of trace metal ions. A 5000 mL of sample volume containing 10 mg L^{-1} of Hg(II) was passed through the column under optimum conditions to study the breakthrough curve. The fractions of effluent were collected at certain time intervals and analyzed by ICP-OES. Figure 8 shows the breakthrough curves for the analyte ion. The breakthrough volumes for Hg(II) at which the analyte concentration is about 3–5% of initial metal concentration were found to be 4000 mL. The breakthrough capacity obtained is very close to the column adsorption capacity, suggesting the potential application of PANI–MoS$_2$ adsorbent for continuous column operation.

Table 1. Analytical data of preconcentration and breakthrough studies (column parameters: sample pH 6; flow rate 8 mL min^{-1}; eluent vol. 5 mL; sorbent amount 0.25 g).

		Preconcentration Studies				Breakthrough Studies	
Sample Volume (mL)	Hg(II) Amount (µg L^{-1})	E(%) [a]	PL [b] (µg L^{-1})	PF [c]	Column Adsorption Capacity (mg g^{-1})	Breakthrough Volume (mL)	Breakthrough Capacity (mg g^{-1})
1500	0.66	100	0.66	300			
2000	0.50	100	0.50	400	240.0		
2700	0.37	100	0.37	540			
3200	0.31	100	0.31	640		4000	160.5
3500	0.29	90	-	-			
4000	0.25	85	-	-			

[a] Extraction percentage; [b] Preconcentration Limit; [c] Preconcentration Factor.

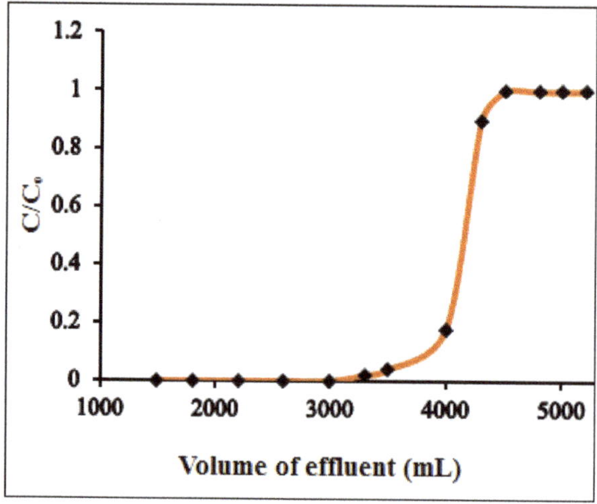

Figure 8. Breakthrough curve for the adsorption of Hg(II) (experimental conditions: sorbent amount 0.5 g; pH = 6.0; flow rate = 8 mL min^{-1}, Hg^{2+} 10 mg L^{-1}).

3.4. Amount of Sorbent and Choice of Eluent and Concentration

The effect of adsorbent dosage on the column preconcentration of Hg(II) was investigated from 0.1 to 1.0 g of the PANI–MoS$_2$-packed column. A model solution of Hg(II) (sample vol. 100 mL; Hg^{2+} = 10 mg L^{-1}) was passed through the column, following the optimized experimental conditions. It was observed that by increasing the adsorbent amount from 0.1 to 0.25 g, the percent recovery of Hg(II) increases and reached 100% at 0.25 g of adsorbent; and remains constant up to 1.0 g of PANI–MoS$_2$ (Figure 9). For subsequent experiments, 0.5 g of adsorbent was optimized for the rest of the experiments. The complete desorption of adsorbed metal ions using a suitable eluent is necessary to reuse the column for the next adsorption cycle. A different eluting agent such as acetic acid, hydrochloric and nitric acids with varying concentration (0.25–1.0 M) and volumes (2–5 mL) was passed through the column with a flow rate of 2 mL min^{-1}. The eluent solution of hydrochloric and nitric acids resulted in the varying recovery of Hg(II) (Figure 10); among them, 5 mL of 0.5 M hydrochloric acid at a flow rate of 2 mL min^{-1} suitably desorbed the Hg(II) (recovery > 99.9%) and prepared the column for next adsorption experiments. Therefore, 5 mL of 0.5 M hydrochloric acid at a flow rate of 2 mL min^{-1} was used as eluent for further experiments.

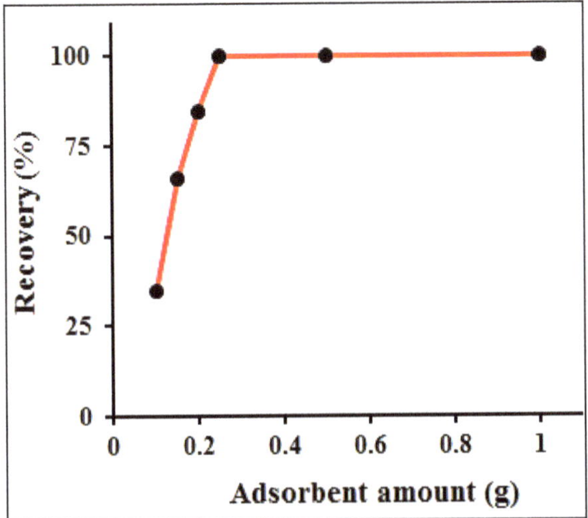

Figure 9. Effect of adsorbent amount on the adsorption of Hg(II) (experimental conditions: sample volume 100 mL; pH = 6.0; flow rate 8 mL min^{-1}; Hg^{2+} 10 mg L^{-1}).

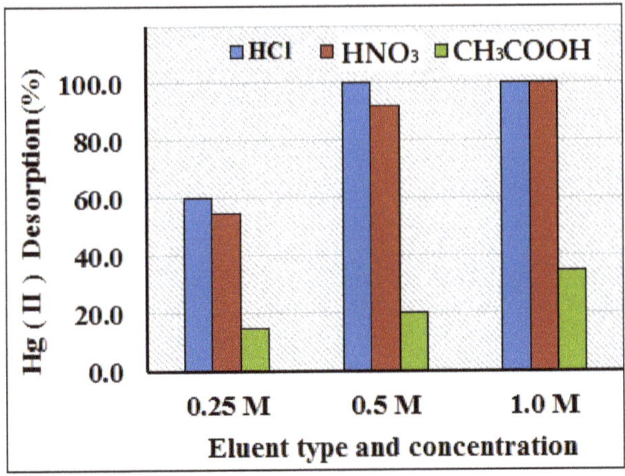

Figure 10. Effect of type and concentration of eluting agents on the desorption of Hg(II) (experimental conditions: sample volume 100 mL; pH = 6.0; flow rate 8 mL min^{-1}; Hg^{2+} 100 mg L^{-1}).

3.5. Influence of Column Flow Rate on Preconcentration Efficiency

The sample flow in the analyte adsorption alters the analyte extraction efficiency and rules the analysis time. Generally, an optimized sample flow permits an equilibrium between the metal ions and the column adsorbent to facilitate the adsorption performance. The effect of flow rate on the adsorption of Hg(II) was investigated by varying sample flow rates from 2 to 10 mL min^{-1} with 100 mL of 10 μg L^{-1} sample solutions at pH 6.0. As shown in Figure 11, the complete recovery of Hg(II) was attained up to a flow of 8 mL min^{-1}. On increasing the sample flow to 9 mL min^{-1}, 92% of Hg(II) recovery was observed due to insufficient contact between the analyte and active sites of PANI–MoS$_2$. Hence, 8 mL min^{-1} of the column flow rate was optimized for the rest of the experiments.

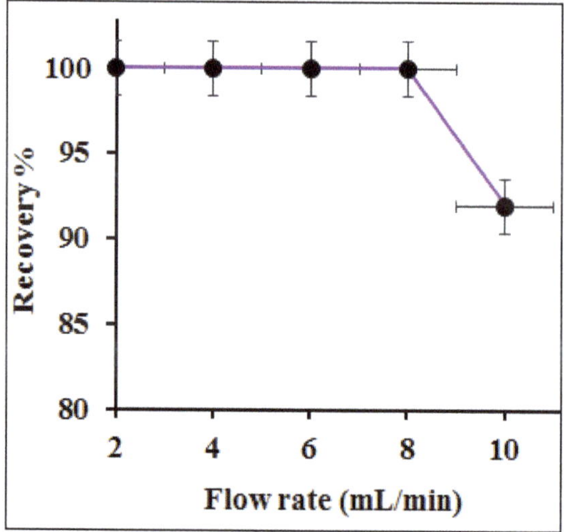

Figure 11. Effect of sample flow on the adsorption of Hg(II) (experimental conditions: sample vol. 100 mL; pH = 6.0; Hg^{2+} 10 mg L^{-1}).

3.6. Interference Studies

The effect of co-existing ions such as ferric, nitrate, carbonate, chloride, sulfate, phosphate and heavy metal ions, including alkali and alkaline earth metal in sorption of Hg(II), were investigated, and the observed data were reported in Table 2. The tolerance level of co-ions was studied by passing a model solution (vol. 100 mL; Hg^{2+} conc. 10 μg L^{-1}) contained a varying concentration of interfering ions through the PANI–MoS$_2$ packed column. The tolerance limit was set as the concentration of co-ions results in a deviation of ±5% in the signal intensity of recovered Hg(II). Under optimum conditions, the proposed method demonstrates fairly good tolerance against co-ions with good recovery of Hg(II) was achieved in the range of 98–100% for quantitative determination.

Table 2. Interference studies on the adsorption of analyte ions (experimental conditions: M^{n+} = 100 μg L^{-1}, sample volume = 100 mL, pH = 6.0, flow rate 8 mL min^{-1}, eluent 5 mL of HCl; N = 3).

Interfering Ions	Salt Added	Amount Added ($\times 10^3$ μg L^{-1})	Recovery % (RSD) Hg(II)
Na$^+$	NaCl	6000	98.0 (4.15)
K$^+$	KCl	5600	98.9 (4.65)
Ca^{2+}	CaCl$_2$	900	97.0 (3.00)
Mg^{2+}	MgCl$_2$	1500	99.7 (4.00)
Cl$^-$	NaCl	9000	100 (4.23)
Br$^-$	NaBr	8000	99.8 (3.54)
CO$_3$$^{2-}$	Na$_2$CO$_3$	4500	98.7 (4.18)
SO$_4$$^{2-}$	Na$_2$SO$_4$	4200	98.6 (4.25)
NO$_3$$^-$	NaNO$_3$	3500	100.4 (4.05)
CH$_3$COO$^-$	CH$_3$COONa	4000	96.5 (4.94)
C$_6$H$_5$O$_7$$^{3-}$	Na$_3$C$_6$H$_5$O$_7$	3300	99.5 (4.16)

3.7. Analytical Figures of Merit and Method Validation

Analytical method validation has been accounted for irrespective of the applicability of the developed procedure for gaining useful data. Following the optimum experimental parameters, the calibration plot for Hg(II) analysis was obtained in the range of 0.2 to 100 µg L^{-1} of Hg(II), with a good correlation coefficient, R^2 = 0.9998. The limit of detection (LOD) and limit of quantification (LOQ), obtained as the concentrations equivalent to three times and ten times of the standard deviation of eleven blank runs, were found as 0.06 µg L^{-1} and 0.2 µg L^{-1}, respectively [37]. Thus, it allows for the ultra-trace determination of Hg(II) in water samples. The relative standard deviation (RSD) that characterizes the method's precision, evaluated for eleven replicate samples containing 5 µg L−1 of Hg(II), was found in the range of 3.0–4.5%. The validity of the proposed method was observed by analyzing the standard reference material (SRM 1641d). The results are shown in Table 3. The closeness of measured value with the certified values is in good agreement, indicates the accuracy of the developed method. In addition, the spiking analysis with two levels of Hg(II) concentration was carried out using different environmental water samples such as household water (tap), industrial wastewater and river water samples (Table S1). The recoveries of the added amount of Hg(II) were satisfactorily recovered with a 95% confidence limit, and the mean percentage recoveries range between 99.0% to 100.2%, with an RSD value in the range 0.35–2.26%. This suggests the accuracy of the method to preconcentrate the trace analytes in real water samples for accurate determination.

Table 3. Analytical method validation by analyzing standard reference material (SRM) after column preconcentration (column conditions: sample volume 100 mL, flow rate 8 mL min^{-1}, eluent 5 mL HCl, sorbent amount 0.5 g).

Samples	Analyte	Certified Values (µg g^{-1})	Values Found by Proposed Method (µg g^{-1}) [a] ± Standard Deviation	Value of *t*-Test [b]
NIST SRM 1641d	Hg(II)	1.56 ± 0.02	1.55 ± 0.06	1.37

[a] Mean value, N = 3; [b] at 95% confidence level.

4. Conclusions

A novel organic–inorganic hybrid adsorbent was synthesized by surface modification of bare MoS$_2$ using PANI. The prepared PANI–MoS$_2$ hybrid material shows selective extraction of Hg(II) in presences of co-existing ions. The fast and selective Hg(II) adsorption may be attributed to the soft acid-soft base interaction between the Hg(II) and–S ions of the PANI–MoS$_2$ adsorbent. A comparative data on the Hg(II) adsorption capacity of prepared material with previous literature was compared and is shown in Table 4. The PANI–MoS$_2$ adsorbent shows comparable adsorption capacity over the previously reported nanoadsorbents. The proposed method's accuracy was validated by analyzing reference material and the standard addition method (RSD < 5%). The proposed methodology is simple and successfully used in the quantitative analyses of trace Hg(II) to monitor the Hg(II) level in real environmental water samples.

Table 4. Hg(II) adsorption capacities of different nanomaterials based on previous literature.

Adsorbent	Metal Ion	Adsorption Capacity (mg g^{-1})	References
PANI–MoS$_2$	Hg(II)	240	This work
MOF	Hg(II)	627.6	[38]
Fe$_3$O$_4$@SiO$_2$SH	Hg(II)	132.0	[39]
MSCFM	Hg(II)	160.4	[12]
Titanate nanoflowers	Hg(II)	454.5	[40]
Magnetic composite	Hg(II)	149.3	[41]

Supplementary Materials: The following are available online at http://www.mdpi.com/2073-4360/12/11/2731/s1, Figure S1: Zeta potential envelope of bare PANI and bare MoS$_2$, Figure S2; ATR-IR spectra of PANI-MoS$_2$ before and after Hg(II) adsorption, Table S1: Solid phase extraction and preconcentration of trace Hg(II) in real samples analyses after to determine Hg(II) concentration by ICP-OES (column conditions: sample volume 250 mL, flow rate 8 mL min^{-1}, eluent 5 mL HCl, sorbent amount 0.25 g).

Author Contributions: Conceptualization, H.A.; methodology, H.A.; software, R.A.K.; validation, H.A.; formal analysis, H.A.; investigation, H.A.; resources, A.A.; data curation, H.A., I.I.B., and R.A.K.; writing—original draft preparation, H.A. and R.A.K.; writing—review and editing, H.A., R.A.K., I.I.B., and A.A.; visualization, H.A.; supervision, A.A.; project administration, R.A.K. and A.A.; funding acquisition, A.A. All authors have read and agreed to the published version of the manuscript.

Funding: The authors extend their appreciation to the Deputyship for Research and Innovation, "Ministry of Education" in Saudi Arabia, for funding this research work through project no. IFKSURG-1438-006.

Conflicts of Interest: The authors declare no conflict of interest.

References

1. Driscoll, C.T.; Mason, R.P.; Chan, H.M.; Jacob, D.J.; Pirrone, N. Mercury as a global pollutant: Sources, pathways, and effects. *Environ. Sci. Technol.* **2013**, *47*, 4967–4983. [CrossRef] [PubMed]

2. Kim, K.-H.; Kabir, E.; Jahan, S.A. A review on the distribution of Hg in the environment and its human health impacts. *J. Hazard. Mater.* **2016**, *306*, 376–385. [CrossRef] [PubMed]

3. Agency for Toxic Substances and Disease Registry. Available online: https://www.atsdr.cdc.gov/spl/index.html#2019spl (accessed on 8 October 2020).

4. Jan, A.T.; Azam, M.; Siddiqui, K.; Ali, A.; Choi, I.; Haq, Q.M.R. Heavy Metals and Human Health: Mechanistic Insight into Toxicity and Counter Defense System of Antioxidants. *Int. J. Mol. Sci.* **2015**, *16*, 29592–29630. [CrossRef] [PubMed]

5. Tchounwou, P.B.; Yedjou, C.G.; Patlolla, A.K.; Sutton, D.J. Heavy metal toxicity and the environment. In *Molecular, Clinical and Environmental Toxicology*; Springer: Basel, Switzerland, 2012; Volume 101, pp. 133–164. [CrossRef]

6. Genchi, G.; Sinicropi, M.S.; Carocci, A.; Lauria, G.; Catalano, A. Mercury Exposure and Heart Diseases. *Int. J. Environ. Res. Public Health* **2017**, *14*, 74. [CrossRef]

7. Lohren, H.; Blagojevic, L.; Fitkau, R.; Ebert, F.; Schildknecht, S.; Leist, M.; Schwerdtle, T. Toxicity of organic and inorganic mercury species in differentiated human neurons and human astrocytes. *J. Trace Elem. Med. Biol.* **2015**, *32*, 200–208. [CrossRef]

8. Environmental Protection Agency. Available online: https://www.epa.gov/ground-water-and-drinking-water/national-primary-drinking-water-regulations#Inorganic (accessed on 8 October 2020).

9. Gómez-Ariza, J.L.; Lorenzo, F.; García-Barrera, T. Comparative study of atomic fluorescence spectroscopy and inductively coupled plasma mass spectrometry for mercury and arsenic multispeciation. *Anal. Bioanal. Chem.* **2005**, *382*, 485–492. [CrossRef]

10. Nakadi, F.V.; Garde, R.; da Veiga, M.A.M.S.; Cruces, J.; Resano, M. A simple and direct atomic absorption spectrometry method for the direct determination of Hg in dried blood spots and dried urine spots prepared using various microsampling devices. *J. Anal. At. Spectrom.* **2020**, *35*, 136–144. [CrossRef]

11. Vicentino, P.D.O.; Brum, D.M.; Cassella, R.J. Development of a method for total Hg determination in oil samples by cold vapor atomic absorption spectrometry after its extraction induced by emulsion breaking. *Talanta* **2015**, *132*, 733–738. [CrossRef]

12. Haseen, U.; Ahmad, H. Preconcentration and Determination of Trace Hg(II) Using a Cellulose Nanofiber Mat Functionalized with MoS2 Nanosheets. *Ind. Eng. Chem. Res.* **2020**, *59*, 3198–3204. [CrossRef]

13. Rastegarifard, F.; Ghanemi, K.; Fallah-Mehrjardi, M. A deep eutectic solvent-based extraction method for fast determination of Hg in marine fish samples by cold vapor atomic absorption spectrometry. *Anal. Methods* **2017**, *9*, 5741–5748. [CrossRef]

14. Ojeda, C.B.; Rojas, F.S. Separation and preconcentration by cloud point extraction procedures for determination of ions: Recent trends and applications. *Microchim. Acta* **2012**, *177*, 1–21. [CrossRef]

15. De Diego, A.; Pécheyran, C.; Tseng, C.M.; Donard, O.F.X. Chapter 12—Cryofocusing for on-line metal and metalloid speciation in the environment. In *Analytical Spectroscopy Library*; Sanz-Medel, A., Ed.; Elsevier: Amsterdam, The Netherlands, 1999; Volume 9, pp. 375–406.

16. Zheng, H.; Hong, J.; Luo, X.; Li, S.; Wang, M.; Yang, B.; Wang, M. Combination of sequential cloud point extraction and hydride generation atomic fluorescence spectrometry for preconcentration and determination of inorganic and methyl mercury in water samples. *Microchem. J.* **2019**, *145*, 806–812. [CrossRef]

17. Ribeiro, A.S.; Vieira, M.A.; Curtius, A.J. Determination of hydride forming elements (As, Sb, Se, Sn) and Hg in environmental reference materials as acid slurries by on-line hydride generation inductively coupled plasma mass spectrometry. *Spectrochim. Acta Part B At. Spectrosc.* **2004**, *59*, 243–253. [CrossRef]

18. Liška, I. Fifty years of solid-phase extraction in water analysis – historical development and overview. *J. Chromatogr. A* **2000**, *885*, 3–16. [CrossRef]

19. Buszewski, B.; Szultka, M. Past, Present, and Future of Solid Phase Extraction: A Review. *Crit. Rev. Anal. Chem.* **2012**, *42*, 198–213. [CrossRef]

20. Wu, Y.; Pang, H.; Liu, Y.; Wang, X.; Yu, S.; Fu, D.; Chen, J.; Wang, X. Environmental remediation of heavy metal ions by novel-nanomaterials: A review. *Environ. Pollut.* **2019**, *246*, 608–620. [CrossRef]

21. Wadhawan, S.; Jain, A.; Nayyar, J.; Mehta, S.K. Role of nanomaterials as adsorbents in heavy metal ion removal from waste water: A review. *J. Water Process Eng.* **2020**, *33*, 101038. [CrossRef]

22. Ahsan, M.A.; Deemer, E.; Fernandez-Delgado, O.; Wang, H.; Curry, M.L.; El-Gendy, A.A.; Noveron, J.C. Fe nanoparticles encapsulated in MOF-derived carbon for the reduction of 4-nitrophenol and methyl orange in water. *Catal. Commun.* **2019**, *130*, 105753. [CrossRef]

23. Ahsan, M.A.; Fernandez-Delgado, O.; Deemer, E.; Wang, H.; El-Gendy, A.A.; Curry, M.L.; Noveron, J.C. Carbonization of Co-BDC MOF results in magnetic C@Co nanoparticles that catalyze the reduction of methyl orange and 4-nitrophenol in water. *J. Mol. Liq.* **2019**, *290*, 111059. [CrossRef]

24. Vasconcelos, I.; Fernandes, C. Magnetic solid phase extraction for determination of drugs in biological matrices. *Trac Trends Anal. Chem.* **2017**, *89*, 41–52. [CrossRef]

25. Herrero Latorre, C.; Álvarez Méndez, J.; Barciela García, J.; García Martín, S.; Peña Crecente, R.M. Carbon nanotubes as solid-phase extraction sorbents prior to atomic spectrometric determination of metal species: A review. *Anal. Chim. Acta* **2012**, *749*, 16–35. [CrossRef] [PubMed]

26. Ahsan, M.A.; Jabbari, V.; Imam, M.A.; Castro, E.; Kim, H.; Curry, M.L.; Valles-Rosales, D.J.; Noveron, J.C. Nanoscale nickel metal organic framework decorated over graphene oxide and carbon nanotubes for water remediation. *Sci. Total Environ.* **2020**, *698*, 134214. [CrossRef] [PubMed]

27. Zhi, L.; Zuo, W.; Chen, F.; Wang, B. 3D MoS2 Composition Aerogels as Chemosensors and Adsorbents for Colorimetric Detection and High-Capacity Adsorption of Hg^{2+}. *Acs Sustain. Chem. Eng.* **2016**, *4*, 3398–3408. [CrossRef]

28. Li, J.-R.; Wang, X.; Yuan, B.; Fu, M.-L.; Cui, H.-J. Robust removal of heavy metals from water by intercalation chalcogenide $[CH_3NH_3]2xMnxSn_3{-}xS_6{\cdot}0.5H_2O$. *Appl. Surf. Sci.* **2014**, *320*, 112–119. [CrossRef]

29. Li, J.-R.; Wang, X.; Yuan, B.; Fu, M.-L. Layered chalcogenide for Cu2+ removal by ion-exchange from wastewater. *J. Mol. Liq.* **2014**, *200*, 205–212. [CrossRef]

30. Bag, S.; Arachchige, I.U.; Kanatzidis, M.G. Aerogels from metal chalcogenides and their emerging unique properties. *J. Mater. Chem.* **2008**, *18*, 3628–3632. [CrossRef]

31. Gao, M.-R.; Xu, Y.-F.; Jiang, J.; Yu, S.-H. Nanostructured metal chalcogenides: Synthesis, modification, and applications in energy conversion and storage devices. *Chem. Soc. Rev.* **2013**, *42*, 2986–3017. [CrossRef]

32. Cheng, W.; Rechberger, F.; Niederberger, M. Three-Dimensional Assembly of Yttrium Oxide Nanosheets into Luminescent Aerogel Monoliths with Outstanding Adsorption Properties. *ACS Nano* **2016**, *10*, 2467–2475. [CrossRef]

33. Wang, L.; Wu, X.-L.; Xu, W.-H.; Huang, X.-J.; Liu, J.-H.; Xu, A.-W. Stable Organic–Inorganic Hybrid of Polyaniline/α-Zirconium Phosphate for Efficient Removal of Organic Pollutants in Water Environment. *ACS Appl. Mater. Interfaces* **2012**, *4*, 2686–2692. [CrossRef]

34. Gao, Y.; Chen, C.; Chen, H.; Zhang, R.; Wang, X. Synthesis of a novel organic–inorganic hybrid of polyaniline/titanium phosphate for Re(vii) removal. *Dalton Trans.* **2015**, *44*, 8917–8925. [CrossRef]

35. Pearson, R.G. The HSAB principle—more quantitative aspects. *Inorg. Chim. Acta* **1995**, *240*, 93–98. [CrossRef]

36. Pearson, R.G. Hard and soft acids and bases, HSAB, part 1: Fundamental principles. *J. Chem. Educ.* **1968**, *45*, 581. [CrossRef]

37. Lara, R.; Cerutti, S.; Salonia, J.A.; Olsina, R.A.; Martinez, L.D. Trace element determination of Argentine wines using ETAAS and USN-ICP-OES. *Food Chem. Toxicol.* **2005**, *43*, 293–297. [CrossRef] [PubMed]

38. Zhang, L.; Zhang, J.; Li, X.; Wang, C.; Yu, A.; Zhang, S.; Ouyang, G.; Cui, Y. Adsorption behavior and mechanism of Hg (II) on a porous core-shell copper hydroxy sulfate@MOF composite. *Appl. Surf. Sci.* **2021**, *538*, 148054. [CrossRef]

39. Wang, Z.; Xu, J.; Hu, Y.; Zhao, H.; Zhou, J.; Liu, Y.; Lou, Z.; Xu, X. Functional nanomaterials: Study on aqueous Hg(II) adsorption by magnetic Fe_3O_4@SiO_2-SH nanoparticles. *J. Taiwan Inst. Chem. Eng.* **2016**, *60*, 394–402. [CrossRef]

40. Liu, W.; Zhao, X.; Wang, T.; Fu, J.; Ni, J. Selective and irreversible adsorption of mercury(ii) from aqueous solution by a flower-like titanate nanomaterial. *J. Mater. Chem. A* **2015**, *3*, 17676–17684. [CrossRef]

41. Wang, X.; Zhang, Z.; Zhao, Y.; Xia, K.; Guo, Y.; Qu, Z.; Bai, R. A mild and facile synthesis of amino functionalized $CoFe_2O_4$@ SiO_2 for Hg (II) removal. *Nanomaterials* **2018**, *8*, 673. [CrossRef]

Publisher's Note: MDPI stays neutral with regard to jurisdictional claims in published maps and institutional affiliations.

MDPI

St. Alban-Anlage 66

4052 Basel

Switzerland

Tel. +41 61 683 77 34

Fax +41 61 302 89 18

www.mdpi.com

Polymers Editorial Office

E-mail: polymers@mdpi.com

www.mdpi.com/journal/polymers